Contents

Introduction

Global Waste is Volume 242 in the ***ISSUES*** series. The aim of the series is to offer current, diverse information about important issues in our world, from a UK perspective.

ABOUT GLOBAL WASTE

When we no longer need or want something, we throw it away. It becomes waste, or rubbish. But, what happens when we do throw something away? Where does it go? Is it recycled? Will it decompose naturally? Unfortunately, most of our waste is simply sent to a landfill, or incinerated – both of which have serious environmental consequences. There are, however, alternative methods of waste disposal, as well as steps that can be taken to prevent waste production. This book explores the issues surrounding waste, from both UK and global perspectives.

OUR SOURCES

Titles in the ***ISSUES*** series are designed to function as educational resource books, providing a balanced overview of a specific subject.

The information in our books is comprised of facts, articles and opinions from many different sources, including:

- Newspaper reports and opinion pieces
- Website factsheets
- Magazine and journal articles
- Statistics and surveys
- Government reports
- Literature from special interest groups

A NOTE ON CRITICAL EVALUATION

Because the information reprinted here is from a number of different sources, readers should bear in mind the origin of the text and whether the source is likely to have a particular bias when presenting information (or when conducting their research). It is hoped that, as you read about the many aspects of the issues explored in this book, you will critically evaluate the information presented.

It is important that you decide whether you are being presented with facts or opinions. Does the writer give a biased or unbiased report? If an opinion is being expressed, do you agree with the writer? Is there potential bias to the 'facts' or statistics behind an article?

ASSIGNMENTS

In the back of this book, you will find a selection of assignments designed to help you engage with the articles you have been reading and to explore your own opinions. Some tasks will take longer than others and there is a mixture of design, writing and research based activities that you can complete alone or in a group.

FURTHER RESEARCH

At the end of each article we have listed its source and a website that you can visit if you would like to conduct your own research. Please remember to critically evaluate any sources that you consult and consider whether the information you are viewing is accurate and unbiased.

Global Waste

Series Editor: Cara Acred

Volume 242

Independence Educational Publishers

First published by Independence Educational Publishers
The Studio, High Green
Great Shelford
Cambridge CB22 5EG
England

British Library Cataloguing in Publication Data

Global waste. -- (Issues ; 242)
1. Refuse and refuse disposal. 2. Recycling (Waste, etc.)
I. Series II. Acred, Cara editor of compilation.
363.7'28-dc23

ISBN-13: 978 1 86168 644 2

Printed in Great Britain

MWL Print Group Ltd

Chapter 1

What is waste?

The problem with waste

What is waste and why does it matter?

Waste or rubbish is what people throw away because they no longer need it or want it. Almost everything we do creates waste and as a society we are currently producing more waste than ever before. We do this at home and at work. The fact that we produce waste, and get rid of it, matters for the following reasons:

- When something is thrown away we lose the natural resources, the energy and the time which have been used to make the product. The vast majority of resources that we use in manufacturing products and providing services cannot be replaced. The use of these resources cannot go on indefinitely – we would run out.
- When something is thrown away we are putting pressure on the environment's ability to cope – in terms of the additional environmental impacts associated with extracting the new resources, manufacturing and distributing the goods, and in terms of the environmental impacts associated with getting rid of our rubbish.
- When something is thrown away we are failing to see it as a resource. It is well understood that what is waste to one person may not be viewed as waste by another. A good example of this is scrap metal which has been recycled for many years. Increasingly people are realising that it makes economic sense as well as environmental sense to use 'waste' rather than just throw it away.

The process of using up the Earth's natural resources to make products which we then throw away, sometimes a very short time later, is not 'sustainable' – in other words, it cannot continue indefinitely.

The way in which we consume materials will affect whether we have a sustainable society that leaves resources available for future generations to use. As consumers and producers, we are central to the concept of sustainability. We need to think about how we can use fewer resources ('get more out of less'), how we can make products last for longer (which means we use less and we throw away less) and how we can do better things with our so-called 'waste' than throw it away. We need to see 'waste' as a 'resource'.

The best way of managing our waste is not to produce it in the first place – waste prevention. After that we can think about reducing the amount of waste we do produce. Then there may be an option to re-use the material. The UK Government has developed this approach to derive a hierarchy of options for managing waste – known as 'the waste hierarchy'.

The 'waste hierarchy'

The waste hierarchy specifies the order of preference for dealing with our wastes – with those towards the top of the list more desirable than those towards the bottom.

The hierarchy is a guide. It does not mean that in all circumstances, at all times, a higher option will be better than a lower option. In most cases a combination of options for managing the different wastes produced at home and at work will be needed. But the hierarchy provides a simple rule of thumb guide to the relative environmental benefits of different options.

The problem we have today is that more of our rubbish is dealt with towards the bottom end of the hierarchy than the top. The challenge is to change our attitudes and our practices so that much more of our waste is dealt with by options towards the top of the hierarchy.

⇨ The above information is reprinted with kind permission from Waste Online. Please visit www.wasteonline.org.uk or www.wastewatch.org.uk for further information.

Stages

Includes

Prevention

Using less material in design and manufacture.

Keeping products for longer; re-use. Using less hazardous material.

Preparing for re-use

Checking, cleaning, repairing, refurbishing whole items or spare parts.

Recycling

Turning waste into a new substance or product.

Includes composting if it meets quality protocols.

Other recovery

Including anaerobic digestion, incineration with energy recovery, gasification and pyrolysis which produce energy (fuels, heat and power) and materials from waste; some backfilling operations.

Disposal

Landfill and incineration without energy recovery.

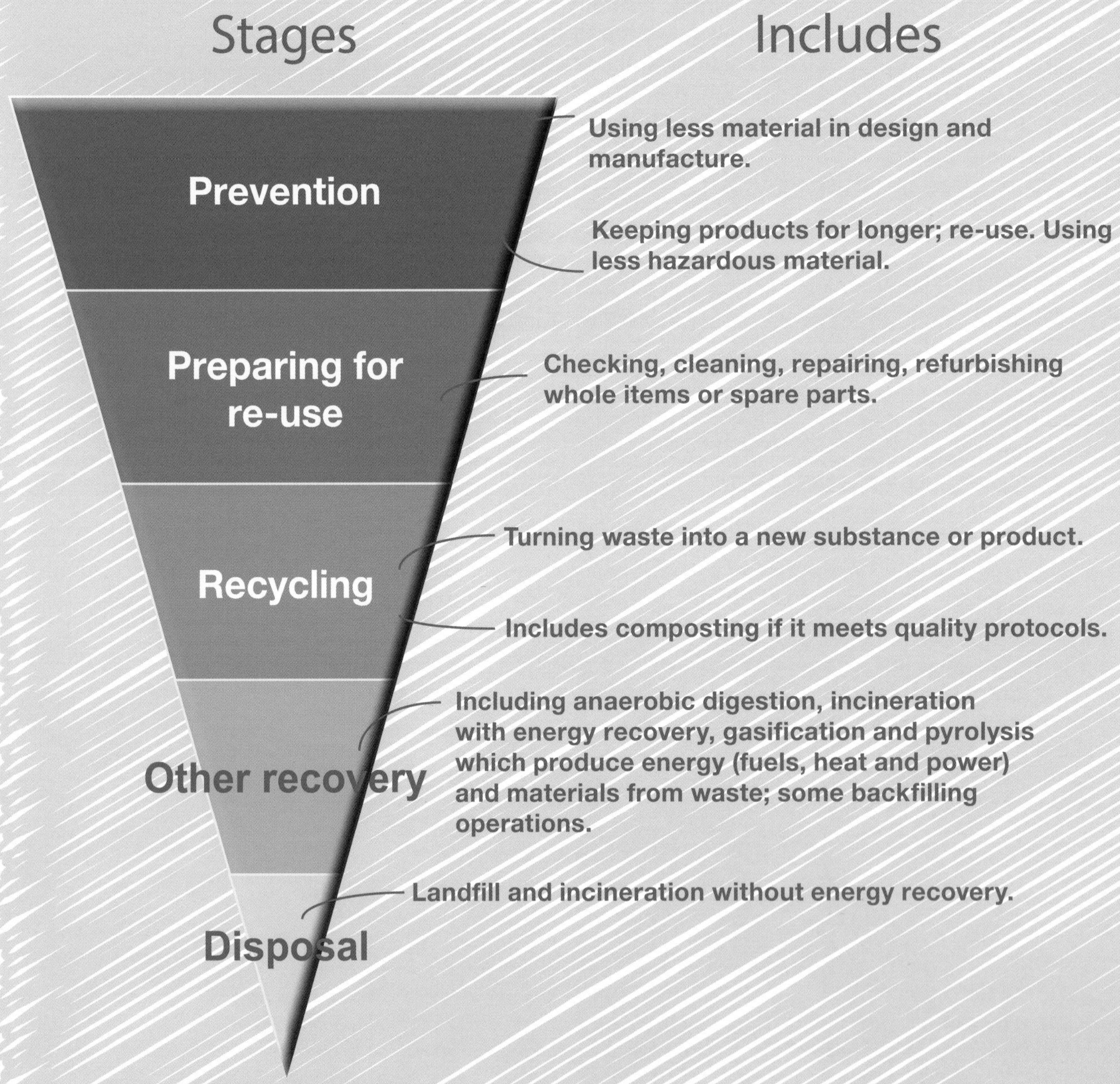

Source: Government Review of Waste Policy 2011. *© Department for Environment, Food & Rural Affairs 2011*

Government review of waste policy 2011

What this means for the environment

Why waste matters

Sustainable waste policies are an important part of tackling climate change and protecting our natural resources. If we want a more sustainable economy, we need to reduce waste and make better use of what we produce. That's why the Government has reviewed and improved its waste policies. This factsheet tells you what we plan to do and how it will help the environment.

A zero waste economy

Despite good progress in reducing waste, the UK is still sending 44 million tonnes a year to landfill. This generates methane emissions and the waste sector is estimated to account for around 3% of all UK emissions. But the overall impact of waste policies on reducing carbon across the economy goes much further. The Government wants to move beyond this throwaway society to a 'zero waste economy' where we re-use and recycle all we can and throw away only as a last resort.

Prevention is best

Our first aim should be to prevent waste by being smarter about how we produce goods and use materials. The Government will help by promoting product design and manufacture that enables easier upgrades, repair and recycling at end of life. These efforts will be targeted at products with high carbon impacts, such as food, metals, plastics, textiles and wood.

Reducing landfill

We will do more to reduce the impact of landfill waste by:

- Consulting on banning wood waste from landfill and considering the feasibility of banning other materials in the future.
- Recovering materials from items such as mobile phones and electrical goods at their end of life.
- Promoting technologies that can capture methane from landfill sites so it can be used as fuel.

Tackling food waste

Every tonne of food and drink wasted creates around four tonnes of CO_2. Food waste also costs business and households money – it can't be right that UK households throw away £12 billion of good food and drink every year. We'll work with the public and private sectors to help businesses and consumers drive down avoidable food waste.

More recycling

We will consult on higher recycling targets for packaging, particularly in areas such as plastic and aluminium where recycling rates are low. You should be able to recycle when you are out and about – we will help businesses and councils who want to work together to put more recycling bins in places like shopping centres and high streets.

Working with businesses

Businesses have a big role to play in managing waste. We will work with them to reduce their waste through voluntary agreements – and invite producers to enter responsibility deals where they take responsibility for the whole life cycle of their products.

What this means for householders

Why waste matters

We all have an interest in waste, from how our bins are collected to the cost of what we throw away – both to us and the environment. That's why the Government has reviewed its waste policies in England – this factsheet tells you about our plans and how they will affect you.

A zero waste economy

We've already achieved a lot – English households recycle 40% of their waste, up from little more than 10% just a decade ago. However, as a nation we still produce 200 million tonnes every year and we need to do more to reduce this. We want to move towards a 'zero waste economy' where we reduce, re-use and recycle all we can, and throw things away only as a last resort.

Better collection services

We want to make it easier for people to re-use and recycle, and easier for people to find out about local waste services. That's why we are encouraging councils to sign up to the new Recycling & Waste Services Commitment. This will tell you what you can expect from your local services and ensure the council listens to you and works with you to help reduce waste.

We understand that householders have a reasonable expectation that waste collection services should be weekly, particularly for smelly waste. We will be working with local councils to increase the frequency and quality of rubbish collections and make it easier to recycle, and to tackle measures which encourage councils specifically to cut the scope of collections.

Recycling on the go

You should be able to recycle when you are out and about as well as at home – we will help businesses and councils who want to work together to put more recycling bins in public places like shopping centres and high streets.

Encouraging re-use

We're working with local authorities, businesses and charities to make it easier for you to donate and buy more re-used items, such as electrical goods, furniture and clothes. For example, we are supporting development of 'one-stop shop' repair and re-use facilities.

Reasonable enforcement

We'll also stop councils from criminalising householders for trivial bin offences such as putting their bin out on the wrong day. Instead, we'll have smaller penalties and target enforcement on those who persistently flout the law, such as fly-tippers. We will support councils that reward and recognise people who do the right thing.

Sensitive planning

In the coming years, landfill sites will close but we will need more sites for re-using and recycling waste, and recovering energy from some types of waste. We'll make sure that householders have a say in planning where these facilities are placed, so that you are involved in the decision that's made.

Find out more

There's more information about waste on these websites:

Defra – www.defra.gov.uk/environment/waste

WRAP – www.wrap.org.uk

Environment Agency – www.environment-agency.gov.uk/business/topics/waste

> ***Did you know?***
>
> ⇨ Anaerobic digestion could produce enough electricity for three million people – the combined populations of Birmingham and Nottingham.
>
> ⇨ Waste services cost the average household over £145 per year.

⇨ The above information is reprinted with kind permission from the Department for Environment, Food & Rural Affairs. Please visit www.defra.gov.uk for further information.

Agenda 21 and waste management

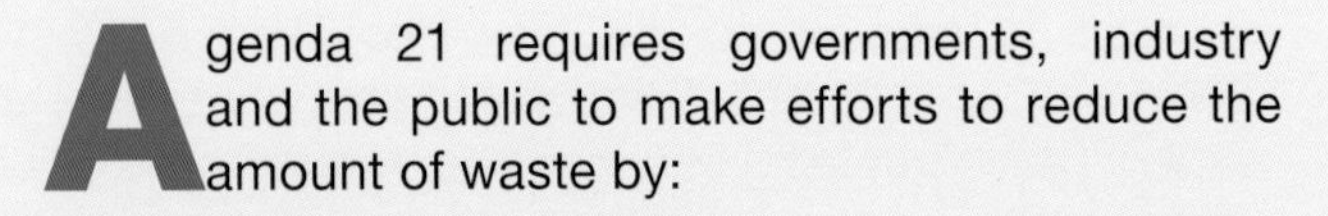

Agenda 21 requires governments, industry and the public to make efforts to reduce the amount of waste by:

⇨ encouraging recycling;

⇨ reducing wasteful packaging of products;

⇨ introducing products that are more environmentally sound.

At the moment, the amount of waste produced in the developed world is not sustainable. For example, 99% of the materials used to make goods in the USA becomes waste within six weeks of sale, including the goods themselves. Most waste is buried underground in landfill sites; in the UK only around 7% of household waste is recycled.

It is estimated that every year in the UK each person throws out the equivalent of:

⇨ Over 100 glass bottles;

⇨ 70 plastic bottles;

⇨ 300 cans;

⇨ 150 newspapers and magazines;

⇨ Over 60 kg of food scraps and kitchen waste.

All of these types of waste can be easily recycled, which saves resources and energy.

⇨ The above information is reprinted with kind permission from Sustainable Environment. Please visit www.sustainable-environment.org.uk for further information.

What happens to our waste?

There are approximately 10,000 landfill sites within the UK. They vary in size and capacity, and some are privately operated and some are owned and operated by the local council.

Many council's are now operating landfill sites in partnership with local companies and land owners. Also, sites vary in their designation to handle only household, non-hazardous waste, or have the facility to deal with hazardous waste, from industry. What is clear is that landfill sites have short lives. For example, it is now commonly known that all the existing landfill sites in East Anglia have a maximum ten years of operation left. It is agreed by all the operators that the UK is producing more annual waste than predicted, and the current landfill sites are filling up faster.

The bottom line is that the country is rapidly running out of land in which to bury all the rubbish we produce. The Government and its agencies have been exporting specific types of waste by sea, chiefly to France and China, but there is mounting pressure against this practice.

At a landfill site: a visit to Dunbar, Scotland

The Viridor Waste Management company operates a large landfill site near Dunbar, on Scotland's east coast. The company is happy for groups to visit and tour the site, and this is highly recommended to get an insight into waste management in the UK.

This site takes at least 150,000 tonnes of waste annually. It is well-served by rail from Edinburgh, the nearest large city. Every morning, a special train arrives bearing rubbish from the city collected the night before. As well as dealing with waste, this site has been at the forefront of recycling rubber tyres by cutting them into chips, and passing them on for further processing.

What is a landfill site?

This site consists of a carefully constructed area, containing a special lining and covering systems to contain the waste. The storage spaces are known as cells, and they are bottomed with layers of clay, sand and then soil. Clay has traditionally been used as a liner, but due to its inclination to crack or fracture, it is now used in combination with materials such as HDPE or PVC plastics.

Unfortunately, despite the best efforts to create effective liners, studies have shown that a ten-acre landfill site will leak between 0.2 and ten gallons of liquid a day through the plastic liner. Some household chemicals found in regular waste, such as alcohol and vinegar, can actually degrade HDPE.

More rubbish, every day

Every day, about 500 tons of rubbish arrives at Dunbar. A JCB compacts the waste, and tips it into the designated cell being filled that day. At the end of each day, the cell is covered with a thick layer of soil. When the cell is full, the final capping layer is put on. This consists of the plastic layer, more soil, a protective geotextile blanket, and a layer of sand. A final soil layer is put on this, which is then planted with vegetation, usually grass.

Under the earth – all our rotting rubbish

In each cell, once these layers have been put on, there is little oxygen and moisture, so decomposition occurs extremely slowly. Bacteria inside each cell breaks down some of the substances to create both a gas and a leachate, which both need to be controlled. The gas has to be vented out of each cell and carefully handled. At Dunbar the gas is methane, which is burnt on site.

The leachate is collected in pipes and pumped up to the surface. It is then recirculated, which has the effect of decreasing its volume, but this also increases the concentration of contaminates within it.

Our rotting, lethal legacy

This is one way to deal with all the rubbish produced in the UK on a daily basis. However, there are reports that a landfill site was excavated 40 years after being covered and closed, and after this time not even the newspapers amongst the rubbish had decomposed!

More serious reports recently found that babies born near to landfill sites were being born with a higher incidence of minor defects. It is still unclear exactly what is causing this, but it is thought to be related to the leachates impacting the local water supply.

All this should be enough to encourage much more intensive efforts both to recycle, and to finding other ways to deal with our mounting rubbish. It is truly an environmental and human problem, and one which we need to find solutions for, urgently.

⇨ The above information is reprinted with kind permission from RecyclingExpert. Please visit www.recyclingexpert.co.uk for further information.

England recycles more than it landfills for first time

More waste was sent for recycling than to landfill in England last year for the first time since records began.

By David Thorpe, news editor

Figures for 2011/12 released by Defra revealed that 43% (10.7 million tonnes) of waste collected by local authorities was sent for recycling, composting or re-use, compared to 37% (9.6 million tonnes) which went to landfill.

'An increase in the amount of waste incinerated may have partly accounted for the change in landfill,' says Defra.

Although this is the highest recycling rate recorded for England, the rate of increase has been levelling off, with 2011/12 being the lowest year-on-year increase for ten years.

This levelling out of improvement could mean that authorities have achieved the easy wins in terms of recycling and are finding it increasingly difficult to influence behaviour change.

However, the fact that some authorities achieve a much higher rate signals that it is possible. The highest rate achieved by a council was 69%, whereas the lowest was 14%.

Defra comments on this disparity by saying that it 'reflects geographical diversity and local waste management infrastructure choices, in part. It can not necessarily be interpreted as a measure of individual authority performance – there is no national framework for local performance measurement.'

Lower rates may reflect urban populations which are difficult to target with individual household recycling bin collections, or rural ones where it is not cost effective to make several journeys with different types of collection vehicle.

Alternatively, lower rates could result from an authority focusing on avoiding landfill by investing in incineration and targeting its waste management policies on that, rather than poor recycling awareness or initiatives.

The European target for household recycling rates is 50% by 2020.

Lord de Mauley, the minister with responsibility for recycling, welcomed the figures, saying: 'Across the country, people are cutting the amount of waste going to landfill by recycling more. They are not only protecting the environment, but fuelling a growing industry that re-uses the things they throw away.

'More still needs to be done and we continue to push towards our aim of a zero waste economy, with businesses, councils and householders all doing their bit.'

The amount of household waste generated is falling year on year, however, at an average rate of 2.6% per year since 2007/08. It was 22.9 million tonnes in 2011/12, representing an average of 431 kg of waste per person.

These statistics also show the amount of greenhouse gases saved from being emitted as a result of recycling, which amounts to 4.26 million tonnes of CO_2-equivalent. By far the most savings come from paper and card, metal, organic wastes and plastic.

Recycling, re-using or composting materials instead of landfilling materials prevented an estimated 6.9 million tonnes CO_2-equivalent. On the minus side, incineration and landfill of waste produced an estimated 2.7 million tonnes of emissions in CO_2-equivalent.

Household waste accounted for 90.1% of all local authority waste managed; the rest is commercial waste, which is estimated infrequently, but is thought to be about twice as much as household waste.

Construction waste generation is even higher, although much is recovered and is inert material that is sometimes recycled into aggregates.

The statistics are based on data submitted by all local authorities in England to WasteDataFlow on the waste they collect and manage, and replace the provisional estimates published for the first three quarters of 2011/12.

The Environmental Services Association (ESA)'s director of policy, Matthew Farrow, also welcomed the figures, saying: 'Our members have been working tirelessly with their Local Authority customers to deliver greater recycling levels, and deserve much credit.'

But he added: 'More needs to be done to reduce the UK's use of landfill. With the Green Investment Bank recently being cleared by the EU, we hope a network of new waste treatment facilities will help move waste up the hierarchy.'

9 November 2012

⇨ The above information is reprinted with kind permission from Energy & Environmental Management. Please visit www.eaem.co.uk for further information.

Call for urgent action over burning and throwing valuable materials in landfill

Government urged to impose bans to stop waste of natural resources as survey finds UK dumps materials worth £650 million.

At least £650 million worth of valuable materials are being thrown into landfills or burned in the UK each year, despite rising costs of natural resources, campaigners and industry warned on Monday.

A coalition of business groups and environmentalists said products ranging from steel, wheat and rubber to rare earth metals necessary for making goods such as mobile phones will become increasingly costly, threatening UK productivity.

The coalition, which includes the manufacturers' organisation EEF and Friends of the Earth, is demanding the Government develop an urgent action plan to preserve valuable resources, including policies to improve recycling and a ban on reusable materials going into landfill.

It comes after a survey by EEF found that four-fifths (80%) of senior manufacturing executives thought limited access to raw materials was already a business risk and a threat to growth, and for one in three companies it was considered the top risk. Last year, the EU's commissioner for the environment told *The Guardian* that the waste of valuable natural resources threatens to result in a fresh economic crisis.

The groups warned the cost of raw materials had surged in recent years, with increases in prices expected to escalate as three billion people join the middle classes across the world, demanding more consumer goods and putting huge pressure on already overstretched natural resources.

But hundreds of millions of pounds worth of reusable materials were being buried in landfills or burned in power plants that generate energy from waste, they said.

The groups urged the Government to ensure that resources are used more efficiently, a move which would create thousands of new jobs, boost the economy and protect the environment.

Ministers should create a new 'office of resource management' to co-ordinate Whitehall action on tackling the resource crisis, set up a task-force to review targets and recommend policies to boost recycling and ban recyclable materials from landfills and energy from waste plants.

The existing resource security action plan, published in March, did not go far enough, they warned.

EEF's head of climate and environment policy Gareth Stace said: 'We live in an age where demand for resources is surging with prices increasing and concerns about shortages mounting.

'While the current action plan was a step in the right direction, it currently falls short of meeting the challenges we will face where obtaining new resources will become more difficult and costly.

'Government must now step up its ambitions and produce a wider plan of action that deals with the challenges not just now but in the longer term.

'This is vital not just from an environmental perspective but to ensure the long-term sustainable future for manufacturing and the wider economy.'

Friends of the Earth resource campaigner Julian Kirby warned ministers must take action to 'prevent a growing resource risk becoming a catastrophe for our economy and the environment'.

He said: 'The UK buries and burns at least £650 million a year of valuable materials, wasting billions of pounds of business and public money.'

A spokeswoman for the Department for Environment, Food and Rural Affairs said: 'The resource security action plan we published this year sets our plans – including a new circular economy task force led by the Green Alliance to encourage better ways of keeping materials in supply chains, a competition to come up with new methods of re-using or recycling precious materials, and further work by WRAP (Waste and Resource Action Programme) to better understand the flow of critical materials in the economy.

'We are working with businesses to strengthen our approach to protecting our economy against materials supply risks, and welcome the EEF's contribution.'

20 August 2011

⇨ The above information is reprinted with kind permission from *The Guardian*. Please visit www.guardian.co.uk for further information.

What is fly-tipping?

What is fly-tipping?

Fly-tipping is the 'illegal deposit of any waste onto land or a highway that has no licence to accept it'. Illegal dumps of waste can vary in scale and the type of waste involved. Tipping a mattress, electrical items or a bin bag full of rubbish in the street causes a local nuisance, and tipping household items and small-scale building or garden waste in open spaces reduces their amenity value to the community. At the other end of the scale there is a growing trend for large-scale fly-tipping which involves several truckloads of construction and demolition waste being tipped on a range of different types of land.

The absence of any formal definition of illegal waste disposal is deliberate. According to guidelines produced by the Department for Environment, Food and Rural Affairs (Defra): 'The definition of fly-tipping is a wide one. This is because there is a general recognition by all, including Government ministers that fly-tipping, whether it is a dumped mattress or a lorry-load of construction and demolition waste can be linked to anti-social behaviour, fear of crime and the liveability of an area.'

Fly-tipping is a growing problem that costs local authorities alone almost £74 million a year to clear up (Flycapture data 2008). At its worst, fly-tipping can lead to serious pollution of the environment and harm to human health if hazardous waste is involved. Fly-tipping can also undermine legitimate waste management business activities.

What causes fly-tipping?

There are several reasons why fly-tipping occurs, and they all need to be considered when tackling this issue.

Ignorance leads to a large amount of illegal dumping. Householders and small businesses pass on their waste to companies offering waste disposal services assuming they are legitimate businesses and will rarely check for any official documentation.

This issue is compounded by the fact that the illegal waste companies are often significantly cheaper than their legitimate counterparts.

Indifference and frustration can also lead to fly-tipping. Householders may opt to illegally tip their waste in a lay-by or other convenient place rather than pay the fee charged by their local authority for a bulky item collection, or locate their nearest civic amenity site. Inappropriate collection services or inadequate waste management infrastructure will also contribute to the problem of fly-tipping.

Research by the Jill Dando Institute suggested that there is a strong link between the type and quality of local waste service provision and levels of fly-tipping. The research suggested that local authorities need to tailor services to the needs of particular communities by raising awareness amongst the elderly, translating information, providing free and regular collection schemes for those without access to cars and paying particular attention to waste collection in areas of high-density housing.

Finally, there is an increasingly large proportion of fly-tipped waste arising from organised criminal activity. As restrictions on waste disposal become tighter and disposal costs rise, there is an expanding market for opportunistic criminals who can undercut the prices of legitimate waste companies and reap significant financial rewards. Illegally disposing of waste cuts costs through the avoidance of landfill tax and by reducing the time taken to transport waste from the place at which it was generated to a tipping site. Legal operators must also charge more because they have invested in training, infrastructure and documentation to comply with legislation.

What items are commonly fly-tipped?

The most commonly fly-tipped materials include:

- ⇨ Components of vehicles: vehicle parts, tyres, oil, petrol
- ⇨ Hazardous waste: asbestos, chemicals
- ⇨ Builders' waste: rubble, bricks, bathroom suites, kitchen units
- ⇨ Trade waste: paperwork, industrial waste
- ⇨ White goods: fridges, freezers, dishwashers, washing machines
- ⇨ Furniture: mattresses, sofas, tables, beds
- ⇨ Garden waste: grass cuttings, shrubs.

Who deals with fly-tipping?

The Environment Agency and local authorities are responsible for removing fly-tipping on public land. Generally the Environment Agency is responsible for removing hazardous waste and large deposits of fly-tipping and the local authority is responsible for removing all other fly-tipping.

Clearing illegally dumped waste from privately owned land is particularly difficult. Neither the local authority nor the Environment Agency is under any legal obligation to remove the waste. Placing a duty on authorities to remove waste from private land would simply encourage illegal dumping rather than tackle the problem. People would not pay the costs of legitimate disposal if they knew they could fly-tip it in the nearest field and the local tax payer would foot the clearance bill.

Tackling fly-tipping and wider waste crime is a priority for the Government. The Government published its waste strategy for England 2007 in May, which set out a blueprint for not only recycling and re-using waste and diverting it from landfill but also preventing it in the first place. The strategy makes it clear that initiatives to boost recycling should be supported by fly-tipping strategies aimed at preventing the illegal dumping of waste.

The strategy has five strands:

- ⇨ Ensuring better prevention, detection and enforcement of the law against fly-tipping and other forms of illegal waste dumping. The Government believes that more effort spent on these aspects will mean less needs to be spent on clear-up and will result in cost savings.
- ⇨ Making existing legislation more usable and effective.
- ⇨ Extending the range of powers available so that the Agency and local authorities can be more flexible when dealing with fly-tipping.
- ⇨ Improving the data and knowledge base so that resources can be targeted.
- ⇨ Ensuring the Agency and local authorities can do their job as effectively as possible and ensuring that waste producers take responsibility for having their waste legally managed.

- ⇨ The above information is reprinted with kind permission from Keep Britain Tidy. Please visit www.keepbritaintidy.org for further information.

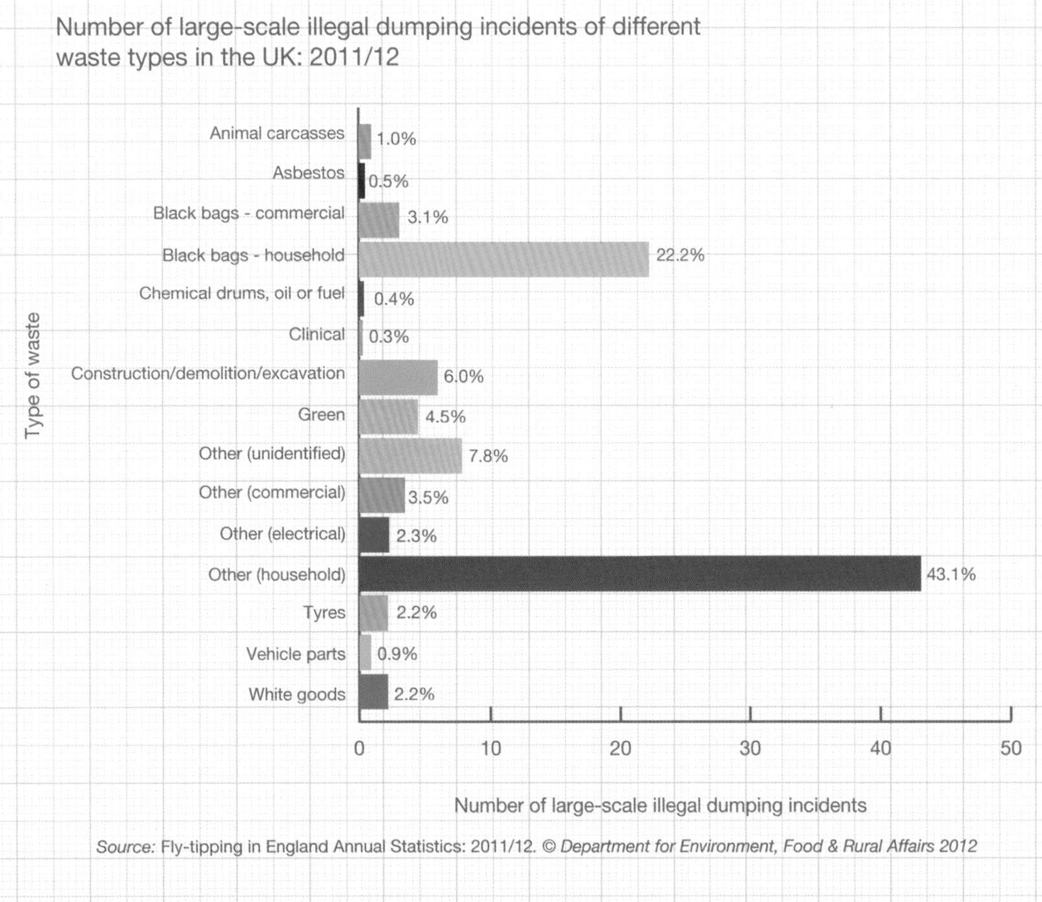

Source: Fly-tipping in England Annual Statistics: 2011/12. *© Department for Environment, Food & Rural Affairs 2012*

Cracking down on waste crime

Waste crime report 2011–2012.

The waste crime problem

What is waste crime?

Waste crime is the deliberate breaking of the law by people who don't manage, transport and dispose of waste correctly.

Not dealing with waste legally can:

- Cause serious damage to the environment, for example illegal waste operations pollute land and rivers.
- Pose risks to human health, for example illegal burning produces toxic fumes.
- Create problems for local communities, such as noise and nuisance for those who live nearby, and declining property prices.
- Be bad for business, as illegal operators undercut legitimate waste companies.

What happens where you live?

Waste crime affects both rural and urban areas in England and Wales. There are crime clusters around areas of higher population density and key motorway links.

The risks of waste crime

Our focus is on tackling serious waste crime which includes illegal waste sites, the illegal export of waste and large-scale illegal dumping.

Illegal waste sites

Illegal waste sites are a top priority for us as they pose a risk to the environment and people. These sites operate without the right physical and operational controls and without the right permit. Sites are also illegal if they have registered an exemption but don't meet its requirements.

Illegal waste activity can also take place on sites with an environmental permit. We are looking at the causes of waste crime to understand more about offender behaviour and its impacts.

Illegal export of waste

Illegal exports of waste move environmental and health and safety risks to countries which often have much less capacity to deal with them than this country.

It is always illegal to export waste from England and Wales for disposal, but some wastes can be exported to certain countries for recycling and recovery.

We must ensure all waste exported from England and Wales meets legal controls and doesn't end up in countries which do not want it.

Illegal dumping of waste

It is a crime to dump waste just anywhere. Dumping incidents can vary significantly in scale. We deal with the large, serious and organised incidents – those that are 'big, bad and nasty'. This sometimes involves hazardous waste which has a greater potential to damage the environment and harm people. Local authorities deal with smaller incidents.

Following a hazardous waste trail

In 2009 we found several trailers containing hazardous waste across the Midlands and north of England. We were concerned about the risk of environmental damage from the waste.

Our investigations led us to a husband and wife, who had already received a fine of £23,000 in 2004 for illegally dumping containers of waste.

Their Preston site had a permit to store up to ten tonnes of hazardous waste. On one occasion we identified the site was holding as much as 250 tonnes.

We worked alongside the local police and their forensic teams to build up an intelligence picture and allow our specialist prosecutors to take action against the pair and two other people who worked with them. The husband was sent to prison for 18 months. The other offenders were ordered to do 120 hours of unpaid work each, and two also received suspended sentences.

Our actions have stopped their illegal dumping and cleared up a site which was significantly exceeding its permitted levels of hazardous waste.

We are also pursuing a Proceeds of Crime Act (POCA) order against those involved in the business to ensure that they do not profit from their criminal activity.

Extent of waste crime

Waste crime undermines legitimate business and the investment and economic growth that goes with it.

About illegal waste sites

Of the 1,175 illegal waste sites we knew about at the end of March 2012 the largest number were for construction and demolition waste (32 per cent). There were a significant number of sites involving mixed household or commercial waste (23 per cent) and end-of-life vehicles and vehicle parts (22 per cent).

Illegal waste sites can store, treat or dispose of waste. Over a quarter (26 per cent) of the illegal sites we knew about at the end of March 2012 were storing waste. We also found a large number (21 per cent) of illegal waste transfer stations.

Focus on reducing illegal waste sites

In 2011–2012 we stopped illegal activities at 759 sites. We did this either by closing them down (670 sites) or helping the site to move into legal compliance and get the right permit (28 sites) or exemption (61 sites) to operate.

Even though we shut down a significant number of sites each year there has only been a slight downward trend in the total number of illegal waste sites. This highlights the sheer scale of the illegal waste site problem and the large number of sites we find each year.

Our challenge is to develop our understanding of this problem further. This is a priority for our new taskforce tackling illegal waste sites.

Since we launched our taskforce in December 2011, the number of illegal waste sites identified has almost doubled. We expected this rise as we developed our intelligence on sites and illegal operators.

The taskforce will continue to collect information to help us understand why sites operate illegally and work out the best way to reduce their numbers effectively. We can then share best practice and use this not only to close down more sites more quickly, but also to reduce the risk of them reopening.

About the illegal export of waste

One of the biggest challenges we face is knowing how much waste is exported illegally from England and Wales. Economic growth and globalisation have seen a worldwide increase in waste exports and imports. This growth has generated a lucrative market in exporting waste, some of which doesn't meet the rules.

Between April 2008 and March 2011 we took part in a Government-funded project to develop and trial intelligence-led inspection, compliance and enforcement approaches to control waste exports. We now only inspect containers that we suspect are non-compliant. As a result, we inspect far fewer containers than we used to do under the old system of random dockside inspections. We carried out this work with partners, including the police, Her Majesty's Revenue and Customs, the UK Border Agency, local authorities and key waste sector representatives.

We have also worked with shipping lines so they can:

- refuse to accept bookings from sites which are known to have made illegal exports
- tell their customers what they will and won't accept in terms of waste shipments, particularly

waste electrical and electronic equipment (WEEE) shipments.

We have also worked with the police on site raids and arrests when illegal exporters are also involved in other serious and organised crime.

Serious charges for the illegal export of waste

While it is legal to export waste, such as plastics, for recycling in some countries, it is illegal to export waste for disposal. The last thing we want is waste from England and Wales being dumped overseas causing harm to people and the environment. The waste criminals also avoid the costs of recycling in the UK and undermine law-abiding businesses.

We undertook a rigorous two-year investigation into the export of approximately 750 tonnes of contaminated waste to Brazil.

The shipment, described as mixed plastics, was rejected by the Brazilian authorities and the containers were repatriated to the UK. The forensic examination of each of the 89 containers was a painstaking process which lasted several months.

Four individuals and two companies have been charged with environmental

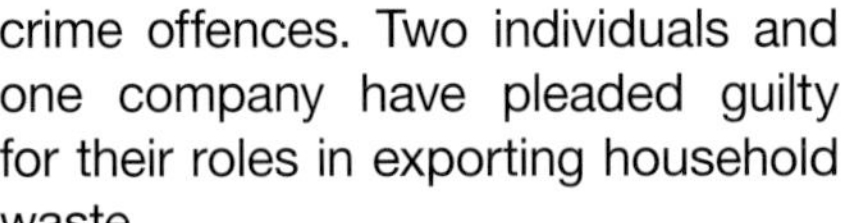

crime offences. Two individuals and one company have pleaded guilty for their roles in exporting household waste.

The case will be heard at the Old Bailey in the autumn, rather than in a magistrate's court, showing the scale and complexity of the case.

About illegal dumping

In 2011–2012 we dealt with 262 incidents of large-scale illegal dumping of waste.

Local authorities in England and Wales dealt with just over 860,000 incidents of smaller-scale and less organised illegal tipping of waste, known as fly-tipping, during 2010–2011.

Our data suggests a significant change in the number of incidents we are dealing with but this is mostly due to the changes in the way we report illegal dumping incidents. Since 2011–2012 we have included dumping that:

- risks pollution of rivers or lakes
- poses an immediate risk to human health, or
- causes a serious risk of flooding.

When the increase due to this change is taken into account, the numbers are broadly similar.

What's being dumped?

Almost a quarter (24 per cent) of illegal dumping incidents that we dealt with in 2011–2012 involved waste from construction, demolition and excavation activities. This is an increase of five per cent from 2010–2011.

We also saw a significant number of incidents involving chemical drums, oil or fuel (19 per cent), and tyres (16 per cent). There was one incident involving clinical waste and one incident involving white goods.

Joining forces to tackle illegal dumping of waste

After our enforcement officers found over 3,000 tyres at an illegal site in Lincolnshire we started to investigate the activities of the so-called 'million-tyre man'.

It currently costs between 60p and 80p to dispose of a tyre in the UK. To avoid paying this much an individual dumped over 800,000 tyres at environmentally-sensitive locations in Essex, Norfolk, Yorkshire, Worcestershire and Lincolnshire. His activities also had an adverse impact on legitimate businesses.

Throughout the investigation we worked closely with the local authorities, fire and rescue services and land agents. With their help we were able to develop intelligence on the disposal route from the point of collection to the multiple illegal waste sites around the country.

We were able to build up a national picture of his illegal activities. This helped our prosecutors bring together his many different offences before the Crown Court to demonstrate the full extent of his offending behaviour.

In November 2011 the courts jailed this serial offender for a total of 15 months.

- The above article contains Environment Agency information and is reprinted with permission. Please visit www.environment-agency.gov.uk for further information.

How clean is England?

The state of England's local environment 2012.

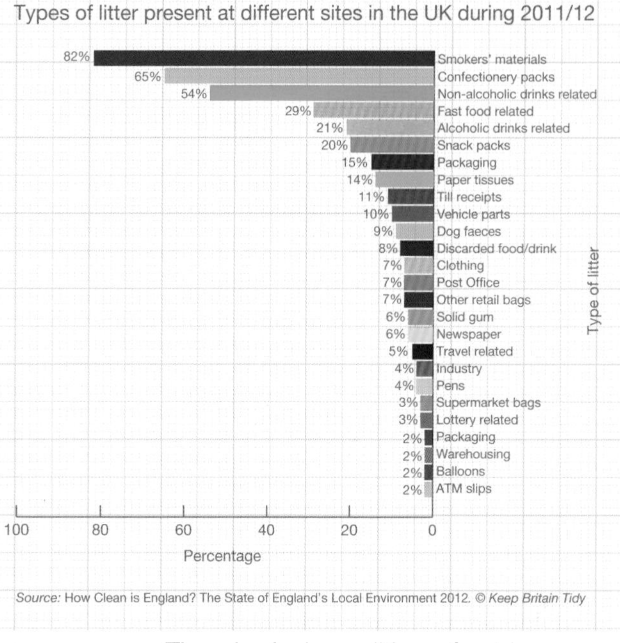

Source: How Clean is England? The State of England's Local Environment 2012. © *Keep Britain Tidy*

What is litter?

Litter includes materials often associated with smoking, eating and drinking, which are improperly discarded and left by members of the public, or items that are spilt during waste management operations. Litter may also include faeces such as dog, bird and other animal faeces.

The survey measures the incidence of different types of litter present in public spaces, such as smoking related, confectionery litter and dog fouling. The proportions of litter from different sources, such as general litter (dropped by a member of the public) and domestic litter (dropped as part of refuse collections, or otherwise emanating from a domestic source), are also recorded.

The maintenance of an area (or lack of) can have an effect on litter levels. Litter in landscaped areas was noted on several occasions by surveyors, in particular mown litter. This indicates that litter is not being cleared before grounds maintenance activities are occurring.

What types of litter are affecting our streets?

During 2011/12 many different types of litter were found. The graph above shows the percentage of sites where the different types of litter found to be present.

Smokers' materials were the most frequently recorded litter type. Cigarette butts and other litter from smoking were found on 82% of sites, a significant 8,807 sites out of a total of 10,725 sites surveyed in England.

Confectionery packs were the second most prevalent item and were found on 65% of sites surveyed. Although present on a large percentage of sites, both these litter items show a decrease in presence from 2010/11 to 2011/12. The third most prevalent was non-alcoholic drinks litter, appearing on 54% of sites.

Fast food litter was the fourth most prevalent, appearing on 29% of sites surveyed, Alcoholic drinks-related was fifth most prevalent, present on 21% of sites and snack packs the sixth most prevalent appearing on 20% of sites surveyed. Items within three of the top six most prevalent items relate to eating food 'on the go'. It is important to consider these findings against wider societal changes, such as people increasingly eating 'on the go' and fast food restaurants and convenience stores extending their opening hours with some being open 24/7.

Litter bins

Evidence has shown that problems of spillage from litter bins can arise even before they start to overflow. It is important to keep litter bins serviced regularly so that members of the public can deposit their litter. Bins should be clean and in good physical condition, so not to deter people from using them. The cleanliness, condition and degree of fill of public litter bins are surveyed. During this study 3,062 litter bins were surveyed.

If a bin is dirty it can deter people from using it. If a person is deterred from using a bin, this may result in littering. A bin is said to be below an acceptable standard if it has a build up of dirt or grime, or heavy amounts of dirt and grime. The variation in cleanliness shows an increase in bins falling below an acceptable standard. 21% of bins were deemed below an acceptable standard of cleanliness, an increase of five percentage points since 2010/11. 79% of bins surveyed appeared to have light amounts of dirt and grime and no bin surveyed was observed to be clean. Looking at trends in bin cleanliness shows that there has been a steady decline in standards over the years and 2011/12 has the worst standard of litter bin cleanliness since 2002/03.

The physical condition of a bin can have an impact on local environmental quality of an area. A bin with significant damage can deter people from using it or can cause litter to spill out. A bin that is no longer functioning means it cannot be used. Research shows that 1% of bins were considered to be of a nearly new condition and 88% of bins surveyed had normal signs of wear and tear. This means that 89% of bins surveyed were said to be of an acceptable standard. No bins were classed as no longer functioning, but 11% were significantly damaged, this is an increase of four percentage points on 2010/11.

Litter can arise from spillage from litter bins, or if bins are too full, it can stop people from using them which can also result in an increase of littering. Looking at all the bins assessed in the survey, 81% of bins were less than 50% full, which means that there does not appear to be an issue with bin emptying frequencies and there would not be a sudden risk in the bins becoming overfull. Just 2% of the bins surveyed were overfull and would have not been able to have been used. 7% were nearly full so were at risk of overflowing and causing spillage.

⇨ The above information is reprinted with kind permission from Keep Britain Tidy. Please visit www.keepbritaintidy.org for further information.

How much food is wasted in total across the UK?

Food is a valuable resource and yet in the UK about 15 million tonnes of food is thrown away every year. Almost 50% of this comes from our homes.

WRAP, the organisation which brings you Love Food Hate Waste not only raises awareness of the issues and benefits of reducing food waste along with easy everyday solutions but also works to bring about changes to the way food is packaged, labelled and sold which make it easier for us all to buy the right amount of food, and use what we buy.

We all have a part to play in reducing the amount of food and drink we throw away – from farm to fork. Lovefoodhatewaste.com has some great ways to help us make a difference and here are some examples of what others are doing:

- People often say that loaves of bread are too big which is why they throw some away – Warburton's launched a range of 600g loaves in late 2008, and Kingsmill launched the 'Little Big Loaf' in 2009.
- Many of us are confused by date labels on our food. Warburton's removed 'display until' dates from all of their products in 2010, bringing them in line with the rest of the bakery sector, and similarly Hovis introduced new storage guidance in early 2011. Many retailers are now removing display until dates from their products, to make it easier to see the important 'use by' and 'best before' dates.
- The Co-operative and Morrisons now provide storage advice on their free loose produce bags, reminding us that keeping most fruit and veg in the fridge helps it stay fresher for longer.
- Heinz launched an innovative 'Fridge Pack' for baked beans in 2010, which could be kept in the fridge for up to five days after opening, giving us longer to eat the product. (Top tip – you can freeze baked beans if you know you won't be able to eat them in time – freeze within two days of opening and then defrost in the microwave and heat till piping hot.)
- Food getting damaged because it is exposed to the air in the fridge and freezer can be avoided by sealing our food well before storing it. Birds Eye introduced re-closable packs for both frozen peas and fish-fingers, to help us reduce waste.
- M&S and Sainsbury's have updated their guidance on when you can freeze their products, making

Source: Love Food Hate Waste. © Love Food Hate Waste 2013

it clearer that you can freeze suitable foods anytime before the date, not just on the day of purchase.

⇨ Asda reviewed their products and as a result gave us an additional 14,000 days of shelf life across 1,672 products. They also launched new re-closable salad bags to help us waste less of it.

Community groups, charities and councils are also making a real difference. A partnership between Love Food Hate Waste and the Women's Institute (WI) helped participants throw away 50% less food waste. The WI now has funding to work with young parents from disadvantaged backgrounds in England as part of its Let's Cook Local project, which Love Food Hate Waste is supporting. Worcestershire County Council recently ran a project in Worcester which reduced the amount those residents wasted by 15% – saving the public and the council money.

The facts about food waste

Did you know...

- ⇨ Almost 50% of the total amount of food thrown away in the UK comes from our homes. We throw away 7.2 million tonnes of food and drink from our homes every year in the UK, and more than half of this is food and drink we could have eaten.
- ⇨ Wasting this food costs the average household £480 a year, rising to £680 for a family with children, the equivalent of around £50 a month.

Environmental impact

- ⇨ If we all stop wasting food that could have been eaten, the benefit to the planet would be the equivalent of taking one in five cars off the road.
- ⇨ The waste of good food and drink is associated with 4% of the UK's total water footprint.

Why is it wasted?

- ⇨ There are two main reasons why we throw away good food: we cook or prepare too much or we don't use it in time.
- ⇨ The foods we waste the most are fresh vegetables and salad, drink, fresh fruit, and bakery items such as bread and cakes.
- ⇨ We throw away more food from our homes than packaging in the UK every year.

The good news!

- ⇨ Between 2006/07 and 2010 food waste has reduced by around 13%, over one million tonnes... this amount of food would fill Wembley stadium!

Remember...

With rising food prices, Love Food Hate Waste really can help you save money every week.

⇨ The above information is reprinted with kind permission from Love Food Hate Waste. Please visit www.lovefoodhatewaste.com for further information.

Date labels

Best Before

BEST BEFORE 04 MAR

These dates refer to quality rather than food safety. Foods with a 'best before' date should be safe to eat after the 'best before' date, but they may no longer be at their best.

Use-by

USE-BY 25 OCT

These dates refer to safety. Food can be eaten up to the end of this date but not after even if it looks and smells fine. Always follow the storage instructions on packs.

Display until

DISPLAY UNTIL 03 FEB

You can ignore these dates as they are for shop staff not for shoppers.

Did you know?

To extend the life of food beyond its date, freeze it before the date and defrost and use within 24 hours.

Providing eggs are cooked thoroughly, they can be eaten a day or two after their 'best before' date.

UK families waste £270 a year on discarded food

Most families massively underestimate the amount of food they throw away each week, according to new research.

By Rebecca Smithers, consumer affairs correspondent

UK families are wasting £270 a year (£5.20 a week) on discarded food and drink, according to a survey of their kitchen habits.

Most families massively underestimate the amount of food they throw away each week, according to new research.

Despite the economic downturn they admit to buying more than they need, often tempted by supermarkets' 'Buy One Get One Free' and similar offers.

The survey of 2,116 adults, carried out by frozen food giant Birds Eye, found that the average household spends £68 a week on food but that 91% of households with children admit to throwing some of that away.

Vegetables topped the list of the most commonly wasted food group, followed by bread and fruit, and 40% of those polled admitted they felt guilty for wasting food.

The main reason cited for wasting food was buying too much (37%), with 22% doing so because of supermarket offers and multi-buy deals.

Lack of meal planning prior to shopping was another issue, with one in three people admitting to not planning.

Families with children at home proved more savvy, with 37% saying they planned more than before the recession.

The research confirms the fact that a large proportion of Britons (almost 70%) have changed their eating habits as a result of the economic downturn – 47% of families are eating out less, 24% have changed what they eat at mealtimes, for example by buying cheaper food, and 26% try to all eat the same food at mealtimes to keep costs down.

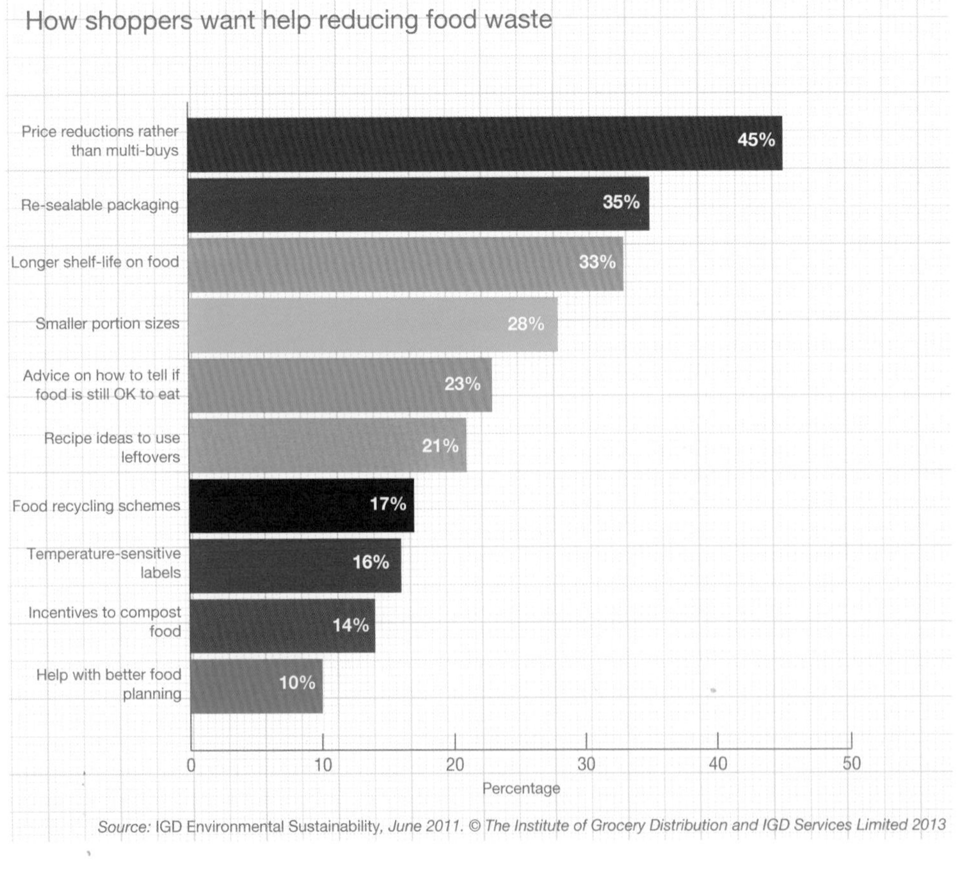

Source: IGD Environmental Sustainability, *June 2011. © The Institute of Grocery Distribution and IGD Services Limited 2013*

The findings come as a new report, *Waste not, want not* by the Fabian Society, which looks at consumer attitudes to food waste, is launched in Parliament on Wednesday. It says that in order to address the problem of food waste, 'it is essential we find fresh ways of communicating about it' and concludes that 'while individuals observe wasteful behaviour in others, they rarely reflect on their own lifestyles as contributing to the problem'.

Waste Minister Lord Taylor of Holbeach said: 'Wasting perfectly good food is bad for household budgets and bad for the environment, which is why we are taking action to help people cut down on what they throw away.

'Through WRAP's Love Food Hate Waste campaign we are helping households to waste less and save money, while our new guidance on food date labelling has cleared up confusion about when food is safe to eat.'

16 May 2012

⇨ The above information is reprinted with kind permission from *The Guardian*. Please visit www.guardian.co.uk for further information.

Our biggest litter count yet!

On 22 March 2013, Keep Britain Tidy released the results of its biggest litter count yet.

And our results show that everyone needs to pitch in and show they love where they live.

For the first time ever, we invited volunteers to join our surveying staff. More than 500 people signed up, allowing us to count more than 37,000 pieces of litter – on the street, in parks, on beaches or beside rivers and canals.

It's a data sample almost more than ten times larger than litter counts in previous years.

Keep Britain Tidy's chief executive Phil Barton said: 'This survey provides us with a snapshot of what people have littered in communities across the country. It also gives 37,000 reasons why we all need to do more to make littering socially unacceptable – to reduce the environmental, social and financial costs of this national problem.'

Mr Barton said while littering is a personal choice that needs to be changed, Government, charities and companies all have important roles in convincing people to make the right choice.

The most littered brand across England was Coke, followed by Cadbury, Walkers, McDonald's and Mars. Red Bull, Imperial Tobacco, Nestlé, Foster's and Pepsi rounded out the top ten.

'These results should be a wake-up call that we all need to do more to love where we live,' Mr Barton continued. 'Litter is not just an environmental problem. It affects perceptions of safety and costs Government nearly a billion pounds a year to clean up.

'Everyone has a role to play, including everyone who buys these products. When you buy a bottle of pop, a bag of crisps or a chocolate bar, you don't just buy the contents you buy the packaging as well and it is your responsibility to dispose of it correctly.'

Mr Barton welcomed the fact that Keep Britain Tidy works with some of the companies whose products feature in this top ten and others, including Coke, McDonald's, Imperial Tobacco, Japan Tobacco International, Wrigley and Mondelez International.

'Some companies and their employees play important roles in changing attitudes and behaviour towards litter through Love Where You Live. It's an effort of Keep Britain Tidy that encourages civic pride – by helping more people and more companies make a real difference in communities.'

Mr Barton noted more will need to be done: 'As Government cuts put pressure on local authority services, and impact on their efforts to get litterers to dispose of things correctly, England could see more litter in more places. That can't happen. It's time for more responsible companies to help more people love more places, by joining efforts to convince their consumers to do the right thing.'

He also thanked the hundreds of volunteers – from community groups to school children – who gave up half an hour of their time to support Keep Britain Tidy's survey.

'We are very grateful to the many volunteers who took time to give us a better idea of what is being littered in England. The strong response shows people do love where they live and want much more to be done to clean up our country.'

In previous years, the count was conducted by Keep Britain Tidy staff, looking purely at town and city centres. This resulted in fast food being the most-littered item by an often-significant margin.

Inviting volunteers to survey more diverse areas resulted in a broader range of most-littered items. The count is not scientific and has always been a snapshot of surveyed areas.

'With much more data this year, one thing is clear. Whether you use, sell or make something, we can all do more to ensure it is disposed of correctly,' Mr Barton concluded.

22 March 2013

TOP TEN WORST BRANDS

1. Coca-Cola
2. Cadbury
3. Walkers
4. McDonald's
5. Mars
6. Red Bull
7. Imperial Tobacco
8. Nestlé
9. Foster's
10. PepsiCo

⇨ The above information is reprinted with kind permission from Keep Britain Tidy. Please visit www.keepbritaintidy.org for further information.

EU exporting more waste, including hazardous waste

Waste is increasingly moving across EU borders, for recovery or disposal. This is true for waste shipments between EU countries, and also transfers of waste outside the EU, according to a new assessment from the European Environment Agency (EEA).

Increasingly stringent and harmonised waste policies in the EU have led countries to transport more waste material elsewhere, for example if they do not have the facilities to recycle or dispose of particular types of waste. There are increasing demands for recyclable materials, both within the EU and beyond, particularly in booming Asian economies.

While trade of hazardous waste grew between 2001 and 2007, shipped volumes decreased in 2008 and 2009, probably due to the economic downturn, according to the report *Movements of waste across the EU's internal and external borders*. Exports of waste plastics and metals picked up again after the economic downturn and exceeded the pre-2009 levels in 2011.

The international trade in recyclable material is expected to continue to grow, the report states, driven by more recycling, growing global competition for resources and increasing awareness of the value of waste. Trade in hazardous waste is also expected to increase, although the driver in this case will be the need to treat waste in specific facilities that are not available in all countries.

Overall the EU should put more efforts into waste prevention in order to become more resource-efficient, a key element of the EU 2020 growth strategy. The report recommends encouraging new technologies and business models that generate less waste, or waste that is less hazardous.

'European countries are exporting more waste than ever,' EEA Executive Director Jacqueline McGlade said. 'The trade in non-hazardous waste can be seen as largely positive, as material is often transported to places where it can be better used. However, we should not lose sight of the bigger picture – in an increasingly resource-constrained world, Europe needs to dramatically reduce the amount of waste it generates in the first place.'

Non-hazardous waste

- Exports of waste iron and steel, and copper, aluminium and nickel from Member States doubled between 1999 and 2011, while waste precious metal exports trebled and waste plastics increased by a factor of five.
- Increasing export volumes and rising prices are both contributing to the growing economic importance of waste exports. The value of scrap iron and steel exports out of the EU has increased by a factor of eight between 1999 and 2011 to €18 billion. Waste copper, aluminium and nickel exports expanded by a factor of six and waste precious metals increased by a factor of 15. The value of annual exports to Asia has grown at an even greater rate.
- Trade in waste wood has also increased steeply. Since 2003, EU imports of waste wood have exceeded exports. Imports of waste wood are primarily driven by the large demand of the particle board industry for wood material. Another demand driver is energy production from solid biomass, which grew by more than 50% between 1995 and 2008.
- Transporting non-hazardous waste for recycling can have positive environmental effects overall, the report notes. Although transporting the material causes additional environmental damage and greenhouse gas emissions, these impacts are often much less than the environmental impacts of processing virgin materials.

Hazardous and electronic waste

- Exports of hazardous waste, which may be explosive, flammable, irritative, toxic or corrosive, grew by 131% in the period 2000–2009, while the amount of hazardous waste generated in the EU increased by 28% in the same period. Flows of hazardous waste into the EU countries, from other EU countries and also from outside the EU, almost trebled between 2001 and 2009, reaching 8.9 million tonnes (Mt).
- Hazardous waste can include fly ash from incinerators, contaminated soil, lead batteries, waste mineral oils and other chemicals. Most hazardous waste exports stay within the EU, going to neighbouring countries. The biggest importer of this material in 2009 was Germany (3 Mt) while the biggest exporter was The Netherlands (2.8 Mt). Most of this material is recycled or used as fuel, although some is still sent to landfills.
- It is illegal to ship hazardous waste from EU Member States to countries which are not members of the Organisation for Economic Cooperation and Development (OECD). Old computers, home appliances and other electronic equipment

should be collected separately under EU legislation.

- However, a large volume of used electrical products are shipped out of the EU to West Africa and Asia, much of them falsely classified as 'used goods' although in reality they are non-functional. The report estimates this trade to be at least 250,000 tonnes every year, possibly much more. These goods may subsequently be processed in dangerous and inefficient conditions, harming the health of local people and damaging the environment.
- The illegal waste trade seems to be growing, the report says, noting that the EU needs to intensify and harmonise inspection activities across the EU to combat illegal waste transfers.

6 November 2012

- The above information is reprinted with kind permission from European Environment Agency. Please visit www.eea.europa.eu

No more plastic waste to China?

Packaging and Films Association warns that UK recycling targets will not be met if China decides to ban plastic waste from the EU.

China is considering a ban on plastic waste from the EU similar to the one recently announced by Malaysia. It is a move that could mark the end of the large-scale importation of plastic waste from the UK and further jeopardise the UK's chance of meeting new Government recycling targets, according to the Packaging and Films Association (PAFA).

'The new recycling targets, already heavily criticised as unrealistic due to the lack of adequate collection and recycling infrastructure, will fail even sooner than expected if these new developments in the Far East come about,' warned Barry Turner, PAFA's CEO.

Turner was referring to an announcement made by China's ministry of environmental protection, ministry of commerce and national development and reform commission, in which it stated it will consider strictly enforcing regulations that prohibit the import of unwashed, post-consumer plastics.

'With much of the 67% of Britain's plastic waste being exported to the Far East, particularly China, according to Defra statistics, and the UK already desperately short of plastic collection and recycling facilities, I believe reaching the target of 57% by 2017 will be even more unrealistic and out of touch,' continued Turner.

However, the news did not come as a surprise to other industry bodies.

'For some time the Chinese authorities have been giving signals that they want to be more reliant on used plastics generated in China.

'The message is that the sooner steps are taken in moving towards a fully fledged domestic recycling system in the UK the better,' Philip Law, Public & Industrial Affairs director, British Plastics Federation told RWW.

This is a view echoed by Dr Adam Read, practice director: resource efficiency & waste management, AEA.

'This is not really a surprise and was always on the cards following what happened in Malaysia. With increasing volumes of locally produced plastic waste the controls on international imports was always going to be under scrutiny.

'We should see this as the 'shock' we need to improve the quality at the kerbside and the MRF and to drive the development of reprocessing facilities in the UK (or in Europe) if that makes more sense,' said Read.

'It underlines the need to keep the quality of recyclate high in order to secure end use markets wherever they are,' Jane Bickerstaffe, director of the Industry Council for Packaging & the Environment (INCPEN) said to RWW.

Matthew Farrow, director of policy, Environmental Services Association (ESA) said: 'Export markets have played a key role in enabling the UK to boost its recycling rates.

'Clearly, however, importers may take a view on what quality specification they are looking for and UK firms exporting recyclates need to meet those. ESA has supported the new packaging targets as ambitious but achievable. It is too early to say whether developments in Asian markets will change this.'

9 October 2012

- The above information is reprinted with kind permission from Recycling & Waste World. Please visit www.recyclingwasteworld.co.uk for further information.

What is e-waste?

Electronic waste (e-waste for short) – also known as Waste from Electrical and Electronic Equipment (WEEE) – covers almost all types of old or broken electrical and electronic equipment (EEE) destined for re-use, recycling or disposal. It includes everyday household appliances such as fridges, washing machines and microwaves, as well as smaller items like coffee grinders, toasters and hair dryers. The fastest growing e-waste is from TVs, computers, phones and related equipment. But it also includes less obvious items such as electronic toys, power tools, smoke detectors, fluorescent lighting and even solar panels.

Categories

The WEEE Directive (2002/96/EC) provides, amongst other things, ten main categories of e-waste. A list and brief summary of the categories is provided below:

Large household appliances:

Fridges, freezers, washing machines, dryers, dishwashers, cookers, microwaves, heaters, radiators, fans and air-conditioning appliances.

Small household appliances:

Vacuum cleaners, irons, toasters, fryers, kettles, scales and other domestic items such as hair dryers, shavers, electric toothbrushes and clocks.

IT and telecommunications equipment:

Computers, laptops, notebooks and accessories (CPU, mouse, screen and keyboard), telephones, printers, copiers, faxes, calculators and other telecommunications equipment.

Consumer equipment:

TVs, radios, DVD players, VCRs, CD players, Hi-Fi items, speakers, amplifiers and musical instruments.

Lighting equipment:

Fluorescent lamps and non-household lighting.

Electrical and electronic tools (except large-scale stationary industrial tools):

Sewing machines, lawnmowers, hedge cutters, strimmers, saws and equipment for turning, sanding, grinding, milling, cutting, shearing, welding and spraying.

Toys, leisure and sports equipment:

Exercise equipment through to computer game consoles, electric train sets and coin slot machines.

Medical devices (except all implanted and infected products):

Dialysis machines, ventilators and radiotherapy equipment and other medical equipment.

Monitoring and control instruments:

Smoke detectors, thermostats and other instruments used in industrial installations, flow gauges and measuring equipment.

Automatic dispensers:

Automatic dispensers for hot drinks, hot or cold bottles or cans, solid products and money.

What are the problems?

E-waste is the fasting growing waste stream in the world. In comparison to other consumer goods, it is growing at an alarming rate due to rapid product innovation and built-in obsolescence. Rapid advances in technology in recent years have produced a multitude of products that we depend on. Increasingly, it's clear that they come with a cost, both to human health and the environment. Here are some of the problems:

Scarcity

To keep pace with growing demand, the scale and speed of resource exploitation for scarce raw materials has accelerated. Many key materials are in danger of running out, whilst mining and extraction on a vast scale is causing unprecedented environmental degradation.

Toxicity

E-waste contains thousands of toxic materials, including heavy metals and harmful persistent chemicals. These can be damaging to both human health and the environment. It's also an impediment to recycling.

Disposability

Explosive sales in consumer electronics and shorter product life-spans means more waste as products are being dumped at an alarming rate. Flatter TV screens, faster laptops and newer multi-functional mobile phones result in technology up-grades long before existing products are worn out. This design for disposability has led to a throwaway consumer culture that is supremely wasteful and environmentally destructive.

Re-use and recyclability

The development of appropriate re-use, recycling and recovery systems is not keeping pace with the sheer quantity of material produced. Huge volumes of usable, workable products are simply discarded or just partially recycled. And, even waste classified as recycled is being illegally exported from Europe to Asia and Africa.

Did you know?

- The UN estimates that some 20 to 50 million tonnes of e-waste are generated in the world each year.
- Global quantities of e-waste are predicted to reach at least 53 million tonnes by 2012.
- Less than 10% of global e-waste is recycled.
- In the European Union (EU27) alone, around 8.7 million tonnes of e-waste is thrown away each year, only 2.1 million tonnes or 25% is collected and treated. The remaining 6.6 million tonnes, or 75%, is hidden and unaccounted for.
- This is predicted to reach about 12.3 million tonnes by 2020.
- The UK has the lowest recycling rate in Western Europe.
- It is estimated that in the UK we throw away around 1.8 million tonnes of e-waste every year. That works out at between 23 and 29 kg of e-waste per head. This compares with a Western European average of between 14 and 24 kg.

- The above information is reprinted with kind permission from E for Good. Please visit www.eforgood.org for further information.

The Great Pacific Garbage Patch

The Great Pacific Garbage Patch is an area of marine debris, laying approximately 135° to 155° West and 35° to 42° North. Although it shifts every year and exact position is hard to tell. It lies within the North Pacific Gyre and does not go anywhere, as it is confined by its currents.

The area

The Patch is around 2,200 kilometres long and 800 audiometers wide (1,760,000 square kilometres) – that's almost three times more than Spain and Portugal combined!

Plastic soup

Consists of both larger and disintegrated plastic objects and particles, both on the surface, in the water column below it and on the bottom.

The 'North Pacific Gyre' (a vortex created by little wind and strong high pressure systems) keeps the soup in constant movement.

Not all plastics float – some (around half of it) are heavier than water and fall to the bottom, affecting the ocean's ecological equilibrium.

The UN Environment Programme estimated recently that each square mile of ocean water contains 46,000 pieces of floating garbage.

How does it form?

Currents in the Pacific Ocean create a circular effect that pulls debris from North America, Asia and the Hawaiian Islands. Then it pushes it into a floating pile of 100 million tonnes of trash.

Where does it all come from?

80% land, brought by sewer systems and rivers to the sea.

20% ships and ocean sources like nets or fishing gear. Also, many containers fall into the sea after severe storms.

Interesting facts

Less than 5% of plastic is recycled.

In the Central North Pacific Gyre, small pieces of plastic outweighed surface zooplankton by a factor of six to one in 1999. But the ratio in 2010 may already be 60 to one.

Photodegradation

Plastic never biodegrades, it doesn't break down into natural substances. But it goes through a photodegradation process, splits into ever smaller and smaller parts, which are still plastic.

Problems created by plastics

- It fouls beaches worldwide and scares tourists away.
- Plastics entangles marine animals and drowns them, strangles them and makes them immovable.
- Plastic litter washed ashore destroys habitats of coastal species.
- Plastic litter gets inside ships propellers and keels, making ship maintenance more expensive.
- Plastic does not biodegrade; plastic things make an ideal vessel and enable invasive species to move to further regions.

- The above information is reprinted with kind permission from the European Commission. Please visit eu.euopra.eu for further information.

Tackling waste

Chapter 2

20 facts about waste and recycling

1. The UK produces more than 100 million tonnes of waste every year; one tonne is about the weight of a small car. In less than two hours, the waste we produce would fill the Albert Hall in London, every eight months it would fill Lake Windermere, the largest and deepest lake in England!

2. On average, each person in the UK throws away their own body weight in rubbish every seven weeks.

3. The average household in the UK produces more than a tonne of waste every year. Put together, this comes to a total of 31 million tonnes per year, equivalent to the weight of three and a half million double-decker buses, a queue of which would go around the world two and a half times.

4. Every year we produce about 3% more waste than the year before. This might not sound like much but, if we carry on at this rate, it means that we will double the amount of waste we produce every 25 years.

5. The average UK family throws away six trees' worth of paper in their household bin a year.

6. Paper and card make up about a fifth of the typical household dustbin. About half of this consists of newspapers and magazines.

7. Two-thirds of paper is recycled, making it one of the main materials recycled in the UK.

8. Each Christmas as much as 83 square kilometres of wrapping paper ends up in UK rubbish bins, enough to cover an area larger than Guernsey, one of the Channel Islands.

9. It is not known how long glass takes to break down, but it is so long that glass made in the Middle East over 3,000 years ago can still be found today.

10. Glass milk bottles are used an average of 13 times before recycling.

11. In 2003, the recycling of glass saved enough energy to launch ten space shuttle missions!

12. We produce and use 20 times more plastic today than we did 50 years ago.

13. Most plastic shopping bags are used only once and a plastic bag can take more than 100 years to decompose!

14. 25 two-litre pop bottles can be recycled into an adult-size fleece jacket.

15. We get through five billion drinks cans every year. Each one could be recycled back into a new can, saving large amounts of energy, raw materials and waste.

16. Weight for weight, empty aluminium cans are worth six to 20 times more than any other used packaging material. There are more than £30 million worth of empty aluminium drinks cans in the UK just waiting to be collected, cashed in and recycled.

17. UK households throw away between £250 and £400 of potentially edible food every year.

18. It has been calculated that, before they are toilet trained, the average child goes through 3,796 nappies, most of which end up buried in landfill sites.

19. It has been estimated that up to 80% of the contents of our dustbins could be easily recycled or composted.

20. Other countries recycle a lot more than we do in the UK. For example, Switzerland, The Netherlands and Germany recycle around 60% of their waste.

⇨ The above information is reprinted with kind permission from C B Environmental Ltd. Please visit www.cbenvironmental.co.uk for further information.

Recycling etiquette

Why is recycling provision so variable across the UK?

The Government does not specify how recycling targets should be met, so it's up to the local authority to implement schemes suited to their area. Services and facilities thus vary greatly, from separated waste collection to the single kerbside 'green box' system. Variation seems endless, and it's due to the following:

- Cost – Investment in new recycling facilities is expensive, so cash-strapped councils stick to established recycling processes (paper, glass).
- Targets – Statutory recycling targets are weight-based, shifting focus onto heavier waste streams (glass, metal) at the expense of lighter plastics.
- Logistics – Collection can be problematic in rural (long distances between homes, scarcity of recycling facilities) and urban areas (limited space, tower blocks).
- No nationwide framework – Industry bodies, charities and campaign groups encourage best practice but there is still a lack of Government guidance.

Which type of collection is best?

Recycling collection schemes aim to, firstly, divert more waste from landfills and, secondly, facilitate efficient, profitable recycling. However, the debate rages on the proper method for meeting these targets:

The case for 'co-mingled' collections

A study by the Waste & Resources Action Programme (WRAP) showed that the quantity of paper collected for recycling rose when collections moved from single-material to multi-material. Clearly, separating recyclables takes time, whereas co-mingled (mixed waste) collections are easier for the householder, and boost overall recycling levels.

To collect the material accepted in co-mingled schemes individually, kerbside collection lorries would need to be highly compartmentalised. Co-mingled kerbside collections reduce the number of trips householders make to recycling centres. Both factors make co-mingled collections more energy-efficient.

The case for separation

Costs increase as more collection and separation is required for the recovery process. Furthermore, co-mingled waste leads to an increased risk of contamination. Different types of material are in contact with each other, and a single kerbside box may result in householders being less attentive when sorting recyclate. The recycling box becomes more of a second dustbin, with hygiene and cross-contamination both issues to be considered.

The solution?

A good compromise is the dual bag method adopted by several local authorities. Powys County Council, for example, provide households with two bags – a red one for plastics and metal, a black one for paper, card and textiles. Partial separation makes the process more efficient for the council, without placing a burden on householders.

How can I improve the way I recycle?

To reduce contamination and improve recycling efficiency, wash and squash!

Wash

- Scrape out any food remains/ pour away excess liquid.
- Rinse the container (use your washing-up water).
- Don't put recyclables in the dishwasher – no need to waste resources to achieve an unnecessary level of cleanliness!

Squash

- Crush metal cans.
- Squeeze plastic bottles flat to expel as much air as possible.

These steps help prevent contamination and reduce the volume of recyclate, making collections more energy efficient.

What about lids, rings and labels?

If you can remove labels and lids from glass jars and bottles, that's great, but don't worry too much because, in the recycling process, the items are re-washed. After crushing, any non-glass objects are removed.

Removing the caps and lids from plastic containers is more important. Plastic caps are often made from a different polymer type, and therefore have a different a melting point when compared to the plastic used for the bottle itself. Too many lids will contaminate the load, so remove and throw away plastic caps where possible. The plastic ring around the neck of the bottle can be left on – a minimal amount of contamination is tolerated.

Remove paper clips, staples and plastic envelope windows from paper. Also remove excessive amounts of tape and labelling from cardboard packaging. Small amounts won't affect the recycling process unduly.

Can I recycle soiled paper?

Paper fibres cannot be recycled if they are contaminated with food. Here are a few tips:

- Put greasy wrappers into your compost/main rubbish.

- Tear out contaminated portions (e.g. a cheesy pizza box lid), and recycle the clean remainder.
- Use tissues as compost, as their dense fibres make them unsuitable for paper recycling.

How do I know what I can recycle?

Check out recycling provision in your area by visiting Recycle Now (www.recyclenow.com). If you are unsure, contact your local authority for details. Ask to speak to the Waste Minimisation Team or Recycling Officer (contact details are included on the Recycle Now website).

Plastics are a particular area of confusion, even though at least two thirds of local authorities now offer a plastics collection scheme. Technically, almost all plastics can be recycled, but the collection infrastructure and low market demand are barriers to the recycling of some types. Blended polymers are particularly costly to recycle, so yoghurt pots, for example, are not usually collected because they are made from a mixture of polymers.

Types of plastic

Almost all plastic bottles are made from one of these plastics:

- PET (bottled for fizzy drinks, cordial, cooking oil)
- HDPE (bottled for milk and fruit juice, washing-up liquid, fabric conditioner)
- PVC (bottles for still mineral water, toiletries, cordial).

Waste Online (www.wasteonline.org.uk) has a detailed list of the common types of plastic and the identifying symbols you will find on the packaging. Alternatively, you can download and print out a handy table to keep by your recycling bin from Recoup (www.recoup.org).

- The above information is reprinted with kind permission from Fubra Limited. Please visit www.recycling-guide.org.uk for further information.

Recycling myths

Recycle for London busts some of the myths that put people off recycling.

'Recycling costs more than sending rubbish to landfills...'

Actually, recycling saves serious money. Every penny spent on landfilling rubbish is a waste of public money.

Check out the Recycling Saves Money section at www.recycleforlondon.com for information on how recycling is cheaper than sending waste to landfill.

'Recycling is too difficult...'

It's easier to recycle than ever before, as councils introduce better collection systems and collect a wider range of materials. Not only is it good for the environment, but it also saves money. If you're confused about what you can recycle where you live, check out our postcode locator.

'Recycling is a waste of time. It all ends up in the same place as the rubbish...'

In London 2010, a whopping 97% of material sent for recycling actually got recycled. The remaining 3% is material that is rejected as it was wrongly placed in the recycling in the first place, either because it can't be recycled or because it is in very bad condition.

'Recycling uses more energy than it saves...'

Using recycled materials to make new products consumes considerably less energy than using raw materials – even when comparing all associated costs, such as transport.

Plus there are extra energy savings because more energy is required to extract, refine, transport and process raw materials ready for industry compared with providing industry-ready materials.

- The above information is reprinted with kind permission from Recycle for London. Please visit www.recycleforlondon.com for further information.

recycle for London

Break the Bag Habit!

Marine Conservation Society (MCS) among leading NGOs joining forces to break England's bag habit.

Join the campaign to bring in a plastic bag charge in England!

MCS has joined The Campaign to Protect Rural England (CPRE), Keep Britain Tidy and Surfers Against Sewage (SAS) to call for a charge on single-use bags in England, following the success of such charges in Wales and Ireland.

The four organisations have formed the Break the Bag Habit campaign calling on Westminster to reduce litter and waste by requiring retailers to introduce a small charge on all single-use bags.

Wales already has one, Northern Ireland is introducing one in 2012 and Scotland is consulting on the matter. Come on England!! Don't get left behind.

How can you help?

You can help put pressure on Westminster to introduce a charge and reduce bag litter in our countryside and in our seas by writing to your MP and/or the media.

Why are we asking for a bag charge in England?

- Over the past two years, the number of carrier bags used in England has increased despite repeated Government calls for retailers to reduce the numbers they give out.
- Last year businesses in the UK issued plastic bags at a rate of 254 a second. A total of eight billion 'thin-gauge' plastic bags were issued during 2011 – a 5.4 per cent increase on the 7.6 billion bags issued in 2010.[1]
- In Wales since the introduction of the 5p charge in October 2011, there have been falls of between 70% and 96% in the number of single-use bags issued. Public support in Wales for this charge has also grown to 70 per cent.[2]
- When Ireland introduced a plastic bag charge in 2002, plastic bag use fell by 90 per cent.[3] Before this charge was introduced, plastic bags made up five per cent of visible litter in Ireland – afterwards, it dropped to 0.32 per cent.[4]
- Last year MCS found 5,433 bags on beaches in one weekend alone accounting for 2.2% of all litter with an average density of 38 bags for every km surveyed.
- In 2011 David Cameron said about this issue: 'Progress overall went backwards last year, and that is unacceptable. Retailers need to do better. I want to see significant falls again. I know that retailers want to do better too but if they don't I will be asking them to explain why not.'[5]

Single-use bags and plastic bags in particular are a menace to the amazing marine wildlife found in English waters. Animals get entangled in them and mistake them for food. This can lead to infections, strangulation, starvation and even death. A charge is a simple, effective way to stop such a pervasive and ubiquitous form of pollution.

- The above information is reprinted with kind permission from Marine Conservation Society. Please visit www.mcsuk.org for further information.

1 WRAP, New figures on carrier bags use, 5 July 2012, http://bit.ly/Mk80ly

2 Welsh Government, Written Statement – 'An update on the single use carrier bags charge in Wales', 4 July 2012, http://bit.ly/LbTJHz

3 Irish Government, 'Plastic Bag Levies: The Irish Experience', 22 November 2011, http://bit.ly/LLoObt

4 See [2]

5 ***Daily Mail*** 29 September 2011, http://www.dailymail.co.uk/news/article-2043014/David-Cameron-warns-supermarkets-Cut-plastic-bags-forced-charge.html

Facts about supermarket carrier bags

Initially the only argument put forward to justify the introduction of levies or bans on carrier bags was to reduce litter, but now it is claimed that levies will also reduce resource consumption and protect marine life.

The litter argument

- There is no correlation between the number of bags used and the number of bags that get littered. The number of littered bags depends on the number of thoughtless or careless people who discard them.
- Carrier bags are NOT a significant cause of litter. According to Keep Britain Tidy's Local Environment Quality Survey (LEQS) reports, supermarket carrier bags were present at only 4% of sites surveyed in 2010/11, down from 7% in 2003/04 – and therefore not present at 96% of sites. It should be noted that even at those 4% of sites where carrier bags were present, there may only have been a single bag.
- Two years after the introduction in 2002 of a levy on plastic carrier bags in Ireland (introduced ostensibly to solve a litter problem) they still constituted 0.22% of litter according to the Irish Litter Monitoring Body. In the latest report, 2010, shopping bags (presumably of all types) are shown as 0.25%. In the UK in 2004 all carrier bags were 0.06% of litter and in 2008 only 0.03%. Source ENCAMS (previous name for Keep Britain Tidy) survey of number of littered items carried out in a sample of the LEQS sites.
- There is no excuse for littering anything in countries with an efficient waste management infrastructure. By contrast, in countries such as Rwanda and Bangladesh with no municipal waste collection services or litter bins, and where bags can clog drains and exacerbate flooding, bans can be justified – though the long-term solution is to introduce proper waste management systems.

Conclusion

A levy on bags which aims to reduce litter in England would be both disproportionate and ineffective.

The resource reduction argument

- A DEFRA/WRAP study IPSOS MORI in 2007 showed that 80% of households re-use their plastic bags at least once for lining bins, wrapping used nappies or food waste before putting them in the bin, or for cleaning up after dogs. Only 6% throw them away immediately.
- The levy on plastic carrier bags in Ireland has had a net negative environmental effect. It resulted in more lorries on the roads (to deliver heavier alternative types of bags to shops) and more plastic being used as sales of tailor-made bags for bin liners and other purposes replaced the carriers. Statistics published by HM Customs and Excise show the amount of all types of plastic bags and sacks used in Ireland in 2001 (before the levy) was 29.8 thousand tonnes and in 2006 it had increased to 31.6 thousand tonnes.

Conclusion

A levy on carrier bags is highly unlikely to conserve total resources (including transport fuel).

The environmental protection argument

- The UK Environment Agency carried out a major study of supermarket carrier bags that concluded a thin plastic bag does least damage to the environment of any carrier bag.
- Plastic bags do not biodegrade in landfill sites. This is a good thing because it means they do not emit greenhouse gases and do not leach chemicals into the water table. And they occupy less than 0.03% of the space in landfills.
- In some areas, where residual waste is collected once a fortnight, people are advised to wrap putrescible waste in carrier bags to guard against flies and rats.

Conclusion

The best way to reduce the (relatively tiny) environmental impact of supermarket carrier bags is to use them more than once.

The protection of the marine environment argument

- It is highly regrettable that some plastic is discarded at sea or finds its way into the oceans. However, media photographs of plastic found on beaches, or recovered from whales' stomachs appear to show they are predominantly large rubbish bags, and bags used in the construction industry, or plastic sheet used in agriculture, rather than carrier bags.
- A 1987 Canadian study in Newfoundland of 100,000 marine mammals and birds attributed their deaths mainly to fishing nets and did not mention plastic bags. This study was misquoted in a 2002 report for the Australian Government which attributed the deaths to plastic bags – only in 2006 was the mistake corrected, replacing 'plastic bags' with 'plastic debris'.

Conclusion

While moves to prevent marine litter are very necessary, the current state of evidence about what that litter is and where it comes from does not support the introduction of a levy on carrier bags.

The 'iconic symbol' argument

- Some policymakers have justified restricting use of bags because they are an iconic symbol of consumer society. If that really is a good reason then it should presumably apply equally to all other symbols such as mobile phones and trainers.
- Some policymakers say that if people take small actions to protect the environment, such as not using carrier bags, this will encourage them to move on to taking bigger actions. However there is no evidence to support this. On the contrary, there is evidence that people feel absolved from doing anything else.

Conclusion

There is no justification for giving carrier bags special treatment and it gives consumers the wrong message to imply that it is a big environmental issue, especially when the Environmental Agency itself concluded that plastic bags were an environmentally responsible way for consumers to carry their groceries home. Policymakers should tackle the big problems, such as energy, food and water shortages and not focus on the trivial.

17 August 2012

- The above information is reprinted with kind permission from Incpen. Please visit www.incpen.org for further information.

Plastic-eating fungi discovered

Could a newly discovered fungus solve the world's plastic waste problems?

A type of fungus that can 'eat' plastic has been discovered in the Amazon rainforest. Students and professors from Yale University found that the fungi can break down a common type of plastic known as polyurethane (PUR).

This is backed up by Mark Osborn, Professor of Microbial Ecology at the University of Hull. He studied plastic that he'd buried in sand on beaches around the UK and found that it could be broken down by certain microorganisms.

These discoveries suggest that fungi could be used to help reduce the problem of plastic waste. However, at present, due to the incredibly large amount of plastic waste in the world, it would take a very long time to break down each piece of plastic in this way.

The plastic industry is concerned that the new discovery could threaten their business. At present, use of the fungus has been controlled but, if it escaped into the environment, it may start to break down plastics that are in use.

Around 260 million tonnes of plastic are produced each year. Much of this ends up in landfill, where it can take hundreds of years to break down, so a way to deal with the problem would be welcome news for the environment.

- The above information is reprinted with kind permission from *First News.* Please visit www.firstnews.co.uk for further information.

Plastic bag ban triggers innovative asphalt

A possible ban on plastic bags caused two brothers to rethink the potential of urban waste.

By Sapna Gopal

Being environmental entrepreneurs was never on the minds of brothers Rasool and Ahmed Khan: they were content running their plastic bag business. Until, that is, a possible ban on plastic bags in Karnataka compelled them to do a rethink. Instead of shutting up shop, they scouted around for viable alternatives.

Intrigued by stories of plastic being used as a constituent of road tar, they began to experiment. With advice from experts at Bangalore University, they started with pothole repair, using a mix of plastic, tar, stones and aggregate. The results were encouraging: the holes stayed filled. Soon their company, now reborn as KK Plastic Waste Management, had won the backing of the Central Roads Research Institute in Delhi, and a patent too.

The process works like this. Plastic bags are shredded, stored for a week to remove moisture, then mixed with asphalt to produce a tough polymerised compound. The resultant substance is stronger than conventional road surfaces, and lasts twice as long (around six years rather than the conventional three) before starting to degrade.

Since plastic has a tendency to act as a binding agent, it increases the ability of bitumen to hold together, even at higher temperatures. (Plastic melts at 130–140°C, bitumen at half that.) And plastic's water resistance helps prevent the roads from becoming waterlogged, even in heavy monsoons. This means they have fewer potholes, so need repairing less frequently than normal surfaces. All this makes the additional 3% construction cost a wise investment.

Thanks to the backing of Karnataka's Chief Minister, S.M. Krishna, the brothers were able to resurface a (highly symbolic) 500 m stretch of road outside the Rajarajeshwari Nagar Gate, and have since covered 1,500 km of the state's roads, using 5,000 tonnes of plastic in the process which would otherwise have gone to landfill – or more likely been burnt.

'People give us their plastic waste – they like the idea that it ends up building their roads.'

They've also boosted the livelihood of traditional kabaadiwalas (waste collectors) and rag pickers, who are paid INR6/kilo for the waste. 'People in apartments and schools in Bangalore even give us things like biscuit packet wrappers and milk packet covers,' adds Rasool Khan. 'They like the idea that what was rubbish ends up in the road.'

Rasool is a strong opponent of banning plastic bags. 'It's just eyewash. It won't be enforced,' he insists. Instead, he advocates collecting plastic from people's doorsteps on a contract basis. This will not only help in the safe disposal of these bags but will also give companies like his the raw material to lay roads successfully. 'If the Government takes the initiative, the problem of disposing plastics can be solved in no time. Banning [all] bags is definitely not a solution.'

18 January 2013

⇨ This article originally appeared in the *Green Futures* Special Edition, 'India: Innovation Nation'. *Green Futures* is the leading magazine on environmental solutions and sustainable futures, published by global sustainability non-profit Forum for the Future. Please visit www.greenfutures.org.uk or www.forumforthefuture.org for further information.

Cleaning up India's waste: but what is the future for army of tip pickers?

New plant may generate power and carbon credits but end hundreds of livelihoods.

By Jason Burke

Rising above Ghazipur, a scruffy suburb in the east of Delhi, sit two gigantic structures. One is a monstrous heap of the city's rubbish that towers over nearby apartment blocks. The second is a new waste processing plant, where the first trials are billed to start next month.

In between lies the local community of waste pickers, more than 400 people who make a living from sorting through the rubbish. They spend their days, often in temperatures well over 40°C, sifting through the mounds of filth as the trucks bring in the waste of one of the world's largest metropolises.

'Our lives are garbage,' says Noor Mohammed, 42, who feeds a family of eight on the 10,000 rupees (£115) he earns each month.

Mohammed has been a waste picker for 20 years. Now, with the incinerator, change is coming which threatens to deprive him of a livelihood. 'I don't know what I will do. We have been told that we will get jobs at the plant. But how can they employ all of us? And we have no skills for other work.'

The project is part of a controversial plan to tackle Delhi's waste problem. A flow of immigrants and economic growth now means 15 million people in the city produce around 8,300 tonnes of rubbish each day, but the city's poor infrastructure, corrupt bureaucracy and poor planning mean the few official facilities are inadequate.

Local officials are optimistic that will start to change with the Ghazipur plant, and two others elsewhere in the capital, which will sift the waste automatically and then burn anything combustible to generate 12 MW of power. 'That will solve the problem of garbage disposal in our area,' says Sajjan Singh Yadav, head of the municipal council of East Delhi.

India, which gets most of its power from burning coal, has applied for carbon credits under the Kyoto protocol for more than 20 similar projects. Hundreds could eventually be built, experts say, with many more worldwide.

The country suffers a chronic lack of energy, with power cuts a feature of daily life. 'The situation is the same not just across India but across the whole developing world. Delhi is at the forefront. It's a test case,' says Federico Demaria, of the Centre for Studies in Science Policy at Delhi's Jawaharlal Nehru University. Earlier this month, the World Bank released a report estimating that city dwellers across the world will generate 2.2 billion tonnes of waste each year by 2025, 70% more than the total today.

But there are as many as two million people who make their living from recycling waste in India. Most walk the few hundred metres from their slum homes at sunrise to the rubbish mountain.

Despite the early hour, trucks had already been driving up the winding road cut into the rubbish to its summit where children clamber through the near vertical stacked filth, searching for plastic or metal which could earn a few coins when sold to intermediaries.

Illness is a way of life. 'We are usually sick five or six days each month,' says Moti Khan, 30, who has been a waste picker for 15 years. 'You get fever, vomiting, diarrhoea, skin rashes, everything.'

The pickers find the detritus of every stage of life: there are often corpses, but also wedding jewellery. No one is sure what will happen to them when the Ghazipur plant opens. Yadav, the official, says there should be about 100 jobs available, though few of these will go the pickers. A few dozen workers would be needed to extract plastic from the refuse but magnets will be used to extract metal waste.

Most jobs will be semi-skilled or technical and, unless they receive training, members of the largely illiterate local community are unlikely to be hired. 'It is up to the contractors who they employ. We are not involved in the welfare of the [waste pickers]. There are some NGOs taking care of it. It is not our responsibility,' he says.

Bharati Chaturvedi, director of Chintan, an NGO that runs a school for the waste pickers, said children would be sent to work elsewhere if the community lost its source of income. The waste pickers and their families live in cramped huts built with material taken from the tip, often 12 to a single room with no sanitation and a single standpipe. Electricity is illegally tapped from a mains cable. A few months ago a power line fell into the slum, causing a fire which destroyed scores of homes.

The scene may be miserable but for most waste pickers it is an improvement on their former living conditions. Most are from Midnapore, a rough, anarchic and desperately poor part of the Indian state of West Bengal. There they were landless labourers, living on the edge of starvation in a region torn by political and social strife. Like so many immigrants to Delhi, what they have now is better than what they left behind.

Some have even managed to save enough to open shops or small businesses and send their children to local schools. One waste picker stood – and lost – in local council elections.

'At least here we get on. It's a life, even if it's garbage. Back there we had nothing,' Nur Mohammed, the veteran waste picker, says. 'This plant will mean an end to all this.'

2 July 2012

⇨ Information from *The Guardian*. Please visit www.guardian.co.uk for further information.

Bin fines up to £1,000 continue

Councils are ignoring the Government by threatening to fine residents £1,000 if they don't recycle.

By Louise Gray, environment correspondent

Eric Pickles, the Communities Secretary, pledged to 'stop the levying of fines by overzealous bin bureaucrats'.

His counterpart in the Department of the Environment has written to all councils asking them to use the fines only as a 'last resort'.

But Colchester Borough Council are introducing a new computer system to make it easier to fine residents.

The new fleet of 13 collection lorries are equipped with a computer system that records how many black sacks are left outside each home in Colchester, Essex.

Householders who put out too much rubbish will be visited by a council warden, who will warn them to start sorting their glass, paper and plastics.

If there's no improvement within three weeks, they'll get a fixed penalty notice, for between £60 and £80.

And if they don't pay, the council could take them to court, where they'll face a fine of up to £1,000.

Last year the Government downgraded the fixed penalty fine from £110 to £80 and promised to stop the 'Talibin'. They also promised to pursue a 'longer term law change' to stop criminal action being taken against well-meaning households.

Martin Hunt, Colchester Borough Council's waste boss, has applied for new powers under the Environmental Protection Act 1990.

The legislation also allows him to order dust carts not to pick up rubbish from serial offenders' homes.

'People are being put off recycling because they can see other people getting away with not doing it,' he said.

'People are saying 'why should I bother with recycling when my neighbour is putting 12 black sacks out?'

'They feel they are paying for someone else not doing what they could be doing, but by me having these powers, it could put a stop to that.'

The new computer system, called Meantime, is being installed in a new fleet of 13 dust carts, which bin men can use to record how many black sacks are being left out by each household on the route.

Householders who are persistently leaving out an unusual amount of black sacks will be visited by a zone warden, who will then monitor the situation for three weeks.

If there is no improvement, the warden will make another visit and continue to monitor collections for another three weeks.

After the second three-week period, if there is no improvement, the resident will be given a fixed penalty notice, for between £60 and £80, which must be paid or challenged.

If the ticket is not paid, the resident's information will be passed on to the courts and the culprits could be fined £1,000.

Local Government Minister, Brandon Lewis, said it was unacceptable to fine people for small mistakes.

'It is completely unacceptable to fine people for the law-abiding activity of disposing of their rubbish in a responsible manner. Such town hall bin bullies are disrespectful to council taxpayers and their petty actions will harm the local environment by encouraging fly-tipping.

'This Government has abolished bin inspectors' powers of entry, scrapped plans for new bin taxes and we will be changing the law to scrap these unfair bin fines.

'This is municipal officialdom gone mad by bureaucrats who are misusing laws passed under the last Government.'

A Defra spokesperson said: 'It isn't right that individuals are treated like criminals if they put out their waste incorrectly. Local authorities will no longer be able to impose criminal penalties on these householders. Local Authorities still need to deal with serious repeat offenders who spoil the local area by the way they put out their waste but this has to be proportionate – they will be able to apply civil sanctions in some cases, but any fines will be lower.'

8 January 2013

⇨ The above information is reprinted with kind permission from *The Telegraph*. Please visit www.telegraph.co.uk for further information.

Rewarding you for recycling

Recycling is easier and more rewarding for a growing number of households in The Royal Borough of Windsor and Maidenhead, thanks to the council's ground-breaking Recyclebank scheme. More than 60,000 households are able to take part.

This scheme follows the hugely successful pilot in 2009 when more than 6,500 households were invited to help test the process, earning Recyclebank points for their recycling efforts.

Since then

- 61% of eligible households have activated their rewards accounts.
- More than 20 million Recyclebank Points have been earned for discounts and offers at over 100 shops, leisure centres, businesses, attractions and cafés/ restaurants; many residents have given their Points to school projects.
- Residents in the trial increased their recycling by an amazing 35%.

The concept is simple – the more you recycle the more rewards you earn

Special blue wheelie bins have been delivered to all appropriate homes and, as residents in the pilot scheme have proved, the new bins make recycling a lot simpler as tins, cans, paper, cardboard, tetra pak, glass bottles and plastic bottles, pots, tubs and trays will go into the one bin – no more separating them into separate black and purple boxes.

To register an account or for more information contact Recyclebank on 0844 409 9490 or online at www.Recyclebank.co.uk.

Community scheme for properties that can't have a blue wheelie bin

Even where a property cannot store or house an individual wheelie bin a new community scheme is now available which means these households do not miss out on rewards for recycling. The scheme means that reward points are allocated based on an equal share of the total amount that was recycled in the area (not including those properties that have their own individual blue bin for recycling). Once residents have told Recyclebank they have recycled, and the area total worked out, points will be deposited into the Recyclebank account of each participating household.

The more your community recycles the more points you earn

You can get fantastic rewards at participating businesses like M&S, Legoland, Magnet and Windsor Leisure Centres or you can donate your points to the Recyclebank Green Schools Programme.

How the blue bin works

The blue bins house a small electronic tag which links to equipment in the collection vehicles where your recycling is weighed to ensure you get the reward points you've earned.

Protecting your privacy

The Royal Borough takes residents' right to privacy seriously. Absolutely no personal details are held on the tag in the wheelie bin. It is only when you choose to take part in the Recyclebank scheme by activating your secure Recyclebank account that the weight of recycling in your blue bin is converted into points for your household so that you can claim the rewards you have earned.

Thanks to residents' efforts the borough's overall recycling rate is already more than 40% – and we believe new rewards scheme will mean we can keep on getting better and better.

More information is available on the Recyclebank website. If you have any questions check the Recyclebank Faqs page or call the Council's Customer Contact Centre on 01628 683801.

- The above information is reprinted with kind permission from The Royal Borough of Windsor and Maidenhead. Please visit www.rbwm.gov.uk or www.recyclebank.co.uk for further information.

Energy from waste

Information from the Institution of Mechanical Engineers.

Letting a resource go up in smoke

As the world becomes more environmentally aware, there is a growing recognition that waste is a valuable commodity. Traditional methods of dealing with waste such as landfilling or burning and burying produce unacceptable harmful emissions. Instead, different waste streams should be regarded as differentiated resources which can be re-used or re-manufactured. For many other types of waste, recovering their value to provide electricity, heat and/or transport fuels is an easy, valuable and more environmentally sound solution than recycling or landfilling.

Energy from Waste (EfW) uses combustion technology and is the only 'renewable' technology which can realistically meet the EU and UK 2020 commitments for 'heat' and 'transport' sector requirements, whilst at the same time also providing significant quantities of electric power. In mainland Europe, recycling and EfW are both used to their optimum potential, and, as a result, landfilling is successfully minimised.

For larger waste streams, combustion technology inherently produces both heat and power, in the ratio of two to three times as much heat energy as electrical, which could make a significant contribution to solving the problem of fuel poverty in the UK.

What are the problems with waste in the UK?

Producing 307 million tonnes of waste (per year) or 'enough to fill the Albert Hall every two hours' is not sustainable. Of that, Defra estimated that 46.4 million tonnes of household and similar waste were produced in the UK with 60% of this landfilled, 34% recycled and 6% used as fuel in EfW plants.

Types of waste

Approximate percentages of waste arising in the UK indicate that around 36% is from construction (high proportion of minerals, plus wood waste); 28% is from mining and quarrying (nearly all mineral waste); 24% is household and commercial (food waste, plastics, metals, glass and paper); 10% is industrial; and around 2% is either agricultural or human sewage. Waste has been regarded as a problem that has to be buried in a landfill. However, with European directives to meet targets in landfill reduction now enshrined in UK law, we believe that there is a real opportunity to avoid landfilling in the future and to regard waste as a resource from which to produce electricity, heat and/or transport fuels.

Resourcing our waste

In an EfW plant, solid wastes such as wood waste from construction and demolition sources are ideally suited for thermal processes. Most plastics have a very high energy content which makes them very suitable for the combustion process. Non-combustibles, such as metals, glass and other inert materials, are unsuitable for EfW plants and are normally recycled by other means.

The myths of recycling

Once 'recyclables' are delivered to the Material Recycling Facility – which is normally a separating and sorting centre – they are officially declared as 'recycled' (i.e. they are counted towards the local or national recycling targets which treat 'sent for recycling' the same as 'recycled'). With very few recycling plants in the UK, many recyclables are actually transported for considerable distances within the country. However, more seriously huge quantities of some major recyclables (particularly paper and plastics), which have already been classified as 'recycled' and counted towards UK and local targets, are being shipped to countries such as China, where we do not know whether they are actually recycled or merely used as cheap fuel.

Going to war!

Waste-as-Resource (WaR) facilities are 'all-in-one' facilities for energy production and recycling. A WaR facility recovers as much energy as possible from the thermal process: a higher proportion of energy normally produced as electricity, but much of the thermal energy (waste heat) is recovered and used in district heating schemes and/or in the various industrial processes in the plant. WaR facilities significantly reduce or eliminate sending ash to landfill, and also incorporate a concrete plant where bottom ash is used as aggregate in a variety of building and construction products.

Denmark: a case study

In most European countries, it is normal to build EfW plants as part of the communities that they serve, so the waste from the community is used as fuel in the EfW plant, which then supplies electricity and heat back to the community.

Denmark was probably the first nation to recognise the resource potential of waste, and has the most notable example of the intelligent use of EfW in Europe. Most, if not all, EfW facilities in Denmark are built close to centres of population, so the waste journey is small and energy produced can be more readily utilised. The electricity produced is used in the local community as is the heat from the thermal process which is distributed in large-scale district heating (DH) systems. The Danes have become world leaders in designing pipelines to deliver heat to buildings over unprecedented distances (over 100 km) with negligible temperature drop.

Is this a wasted opportunity?

The UK has committed to climate change mitigation targets for 2020. Over 90% of the UK's energy supply is provided from fossil fuels, and since these are the biggest single contributor to climate change, it follows that increasing energy demand of whatever form will be largely supplied from fossil fuels and will therefore exacerbate, and not mitigate, climate change. EfW, on the other hand, is utilising a renewable resource as fuel and is, therefore, making a significant contribution to climate change mitigation.

So what needs to be done?

We have a growing pile of waste which needs dealing with and energy production should be a solution. The enormous amount of thermal energy (heat) produced by a combustion process is wasted to the atmosphere, but simple technology can capture much of this and use it for space heating in a district – or community – heating scheme.

A long-term commitment to make use of this energy by developing community heat networks could offer a viable and direct solution to the fuel-poverty issue, alongside much needed and highly cost-effective measures to improve the insulation and thermal efficiency of our existing housing stock. Moreover, a community/regional programme would provide a sustainable economic benefit to construction and engineering companies, could be initially targeted at high fuel-poverty areas and resolve many local waste disposal issues throughout the UK.

Recycling will never be the sole solution for the UK's waste issues – there is quite simply too much waste to deal with and too many waste streams that do not benefit from recycling. Technologies and options are available to segregate the streams which should be recycled from waste that can be used as a valuable and secure energy source. Plus, existing landfills could house EfW plants.

A long-term education programme on the merits of recycling is required to allow EfW plants to be created to both generate energy for local communities and remove large amounts of waste being produced by the same communities. Looking further ahead, full-scale Waste-as-Resource plants would deal with the vast majority of what we currently still think of as 'waste'. Let's not waste the opportunity. The time for Energy-from-Waste is now!

⇨ The above information is reprinted with kind permission from the Institution of Mechanical Engineers. Please visit www.imeche.org for further information.

Why oppose incineration?

Why United Kingdom Without Incineration (UKWIN) seeks a United Kingdom without incineration

The alternatives to incineration are cheaper, more flexible, quicker to implement and better for the environment. Rather than incinerating waste, local authorities should focus on maximising recycling and providing a weekly separate food waste collection for treatment by composting or anaerobic digestion (AD). Recyclables and biodegradables should be separated from the small amount of residue material. This residue should be stabilised by composting and then sent to landfill.

The incineration of household waste

- Depresses recycling and wastes resources
- Releases greenhouse gasses
- Is often forced through against strong public opposition
- Relies on exaggerating future quantities of waste instead of strongly increased recycling and composting
- Creates toxic emissions and hazardous ash
- Poses significant health risks.

Incineration depresses recycling and wastes resources

Incineration reduces our ability to re-use or recycle potentially valuable discarded material. As WRAP notes in its recent report (*Domestic Mixed Plastics Packaging Waste Management Options*), there is currently a 'Catch 22' situation, with few Local Authorities prepared to collect plastic waste other than bottles, as there is limited potential for them to be recycled. However, this means there is a lack of such plastics available for companies to attempt to do so.

WRAP also reports that 'recycling offers more environmental benefits and lower environmental impacts than other waste management options'.

Waste PFI contracts that include incineration depress recycling rates. In Nottinghamshire it would appear that Veolia see it as more profitable to fall short of recycling targets, as their planning application for a waste incinerator to be built in Sherwood Forest indicates that Nottinghamshire's recycling will be effectively capped at less than 47% for the next 25 years!

Incineration releases greenhouse gasses

Incineration involves the release of high levels of CO_2, the main climate warming gas. Accounting for recovered energy, incineration is accompanied by twice or more the CO_2 per unit of power than the same energy (as electricity or combined heat-and-power) produced from fossil fuel (*Stop Trashing the Climate* report, June 2008). The Environment Agency's WRATE software is used to claim energy-from-waste is beneficial, but this depends on faulty assumptions on efficiency and bio-carbon. Proper lifecycle calculations using the better ATROPOS model found that 'scenarios using incineration were amongst the poorest performing' while those using MBT were much better (*Greenhouse Gas Balances of Waste Management Scenarios*, Eunomia Consulting report plus errata to the Greater London Authority, January 2008).

Studies show that for electricity-only incinerators (incinerators that do not optimise the use of the heat they produce), energy production is so inefficient that, from a climate change perspective, incineration is worse than gas- or coal-fired power stations!

Incineration is often forced through against strong public opposition

Incineration is not the way that householders want their discarded material to be managed. Despite this, Defra was given a further £2 billion of funding through PFI credits in the 2007 Comprehensive Spending Review (including £600 million in 2008/09, £700 million in 2009/10 and another £700 million in 2010/11). Defra have issued guidelines for the granting of Waste Private Finance initiative (PFI) funding, but the process is not transparent. Although Local Authorities do not appear to follow Defra's rules, public funding (in the form of PFI credits) is still being made available to them.

For example, Guideline 6 states: 'Proposals should demonstrate that other relevant authorities, the public, and interested parties have been consulted and that there is a broad consensus supporting a recognised long term waste management strategy which is reflected in the proposed solution', yet it would appear that, for instance, Waste Officers from Hull and East Riding Councils submitted an Expression of Interest for funding, and that Waste PFI funding was approved by Defra, without the knowledge of the democratically elected members (Councillors), and certainly without reaching 'a broad consensus' to support their incinerator proposals.

Incineration relies on exaggerating future quantities of waste instead of strongly increased recycling and composting

Many Waste PFI contracts are entered into as a response to predictions of huge increases in the quantity of household waste (often calculated five to ten years ago) when in fact household waste has actually fallen in many areas. According to Defra statistics, the average annual increase in municipal waste from 2001/02 to 2006/07 was just 0.2%, far short of the 3% year on year rises that were predicted. These flawed and exaggerated predictions are still being used to try to justify the building of unnecessary incinerators.

Operators also say they could always 'top up' with commercial and industrial (trade) waste to make up for any 'shortfall', although in practice this has been shown to cause operational efficiency problems, e.g. for Veolia's Sheffield incinerator.

Incineration creates toxic emissions and hazardous ash

While everyone agrees that incinerators do not improve air quality, there is a great deal of controversy over the extent and nature of the harm resulting from releases into the air (and indeed releases to land and water). Although incinerator fumes pass through expensive filter systems, modern incinerators still emit significant levels of NOx and of ultrafine particles. The latter includes nano-particles which are of great concern because they can pass through the lung lining, causing internal inflammation and penetrating to organs (even to the foetus in a pregnant mother) [1].

Dioxins still an issue: officially these most toxic products are restricted to very low emission levels by incinerator filters. But studies overseas show that high levels are emitted during start-up and close-down when dioxins are not monitored in the UK [2].

Incineration poses significant health risks

UKWIN calls for the adoption of a more precautionary approach while better scientific research is conducted into the extent of the damage to human and animal health (and to ecosystems and fragile habitats) caused by the release of these harmful toxins. The scientific evidence is quite sufficient, UKWIN argues, to trigger the precautionary principle. Government and regulators should compel the waste industry to measure, assess and suppress all of their suspect emissions of harmful toxins.

There is plenty of evidence that emissions from incinerators and their ashes are potentially harmful. The licensed emissions of NOx and particulates cause a level of harm that is included in the EU assessments of industrial and traffic emissions. Incinerators also have emissions unlimited by license, during start-up and close-down, and from ash dispersing during transfer to landfill or construction sites [3].

5–7% of the mass of incinerated waste becomes 'fly ash' (also known as APC). The fly ash is trapped by filters, and is classed as hazardous waste. Because fly ash is strongly alkaline and also high in dioxins and heavy metals, it has to be transferred to landfill. The Bishops Cleeve hazardous landfill site in Gloucestershire takes fly ash from many incinerators; the residents see the ash literally blowing around. UKWIN believe residents are fully justified in fearing the health impacts. Indeed, the health risks have been shown as significant by an official study by Duarte-Davidson et al.

Grate-ash (bottom ash) forms another fraction, 25–30% of the mass of incinerated waste. This ash also contains levels of dioxins and metals. Because of their commitment to incineration, the authorities are encouraging the use of bottom ash as construction fill and as an aggregate substitute. But some of this ash spreads around during construction, and the toxins leach into groundwater. During new construction, in decades to come, the metals and dioxins will get into the environment.

⇨ The above information is reprinted with kind permission from the United Kingdom Without Incineration Network (UKWIN). Please visit www.ukwin.org.uk for further information.

Notes

[1] Nano-particles arise in huge numbers from vehicles, most being carbonaceous. But high temperature combustion processes such as incineration generate nano-particles with metallic, dioxin and aromatic hydrocarbon (PAH) coatings, which may be much worse for health. The review by Cormier et al. (*Origin and Health Impacts of Emissions of Toxic By-Products and Fine Particles from Combustion*, 2006) is strong evidence, while various research papers are establishing tangible public health impacts (Univ. of California study 2008 – *Air Pollution May Cause Heart Disease*; shows nano-sized particles are the most damaging). See FOE's workshop paper for wider discussion and specific references.

[2] A 2007 Japanese study (*Characteristics of dioxin emissions at startup and shutdown of MSW incinerators*) implies that dioxins from a few start ups each year can be very significant, while incinerators operated in batch modes would emit a high proportion of the total when starting up. Other studies show dioxin-like products are also emitted, but again these are not controlled or monitored. Also see Peter Montague's article entitled *The Deadliest Air Pollution Isn't Being Regulated or Even Measured*.

[3] As the Royal Society state (in relation to the oft-quoted 2004 Enviros study into the health impacts of incineration): 'In view of the large uncertainties associated with some of the data examined, particularly in the epidemiological studies, it would have been more appropriate to adopt a cautious approach, rather than use inadequate data in a quantitative framework. The latter may give a misleading impression of the robustness of the results. Caveats associated with the uncertainties in the results are not presented adequately, particularly in the quantification of the health effects, which could mislead the reader. The report's relevance to waste management decision-making by Local Authorities is limited, as several important issues are not addressed. These include the effect of local environmental and health sensitivity to pollutants and the impact on emissions of specific waste management activities operating under non-standard conditions. Bias in the availability of good quality information means the report concentrates mainly on the effects of air pollution. Consideration of the potential effects of exposure to pollutants through other pathways is not consistent throughout the report and therefore prevents adequate comparison of the options.'

Sweden runs out of garbage, forced to import from Norway

Sweden, a recycling-happy land where a quarter of a million homes are powered by the incineration of waste, is facing a unique dilemma: the nation has run out of much-needed fuel.

Matt Hickman, eco-living expert

Sweden, birthplace of the Smörgåsbord, Eric Northman and the world's preferred solar-powered purveyor of flat-pack home furnishings, is in a bit of a pickle: the squeaky clean Scandinavian nation of more than 9.5 million has run out of garbage. The landfills have been tapped dry; the rubbish reserves depleted. And although this may seem like a positive – even enviable – predicament for a country to be facing, Sweden has been forced to import trash from neighbouring countries, namely Norway. Sweden is so trash-strapped that officials are shipping it in – 80,000 tonnes of refuse annually, to be exact – from elsewhere.

You see, Swedes are big on recycling. So big in fact that only four per cent of all waste generated in the country is landfilled.

Good for them! However, the population's remarkably pertinacious recycling habits are also a bit of a problem given that the country relies on waste to heat and to provide electricity to hundreds of thousands of homes through a long-standing waste-to-energy incineration programme. So with citizens simply not generating enough burnable waste to power the incinerators, the country has been forced to look elsewhere for fuel. Catarina Ostlund, a senior advisor for the Swedish Environmental Protection Agency says: 'We have more capacity than the production of waste in Sweden and that which is usable for incineration.'

Public Radio International has the whole story (thanks to Ariel Schwartz at Co.Exist), a story that may seem implausible in a country like garbage-bloated America where overflowing landfills are anything but scarce.

As mentioned, the solution – a short-term one, according to Ostlund – has been to import (well, kind of import) waste from Norway. It's kind of a great deal for the Swedes: Norway pays Sweden to take its excess waste, Sweden burns it for heat and electricity, and the ashes remaining from the incineration process, filled with highly polluting dioxins, are returned back to Norway and landfilled.

Ostlund suggests that Norway might not be the perfect partner for a trash import–export scheme, however. 'I hope that we instead will get the waste from Italy or from Romania or Bulgaria or the Baltic countries because they landfill a lot in these countries,' she tells PRI. 'They don't have any incineration plants or recycling plants, so they need to find a solution for their waste.'

25 October 2012

⇨ The above information is reprinted with kind permission from Mother Nature Network. Please visit www.mnn.com for further information.

UK's first 'super' food waste plant opens

On 17 June, 2011, leading UK waste specialist Biffa launched the UK's first 'super' anaerobic digestion (AD) plant dealing with food waste, near Cannock, Staffordshire.

The facility, which is the biggest in the UK, processes up to 120,000 tonnes of food waste from homes and businesses every year to produce enough renewable energy to power 6,000 homes and a soil improver that can be used in the same way as compost.

Speaking at the plant launch, Biffa chief executive Ian Wakelin said: 'This is the future of waste. It is taking food that could once only be sent to landfill and turning it into something of value on a truly industrial scale. It is a key milestone in society's drive to reduce waste, cut emissions and recover the inherent value in our waste.'

The Cannock facility is the latest in a long line of commitments from Biffa to handle more of the UK's food waste. It forms part of Biffa's expanding network of AD plants and follows on from the launch of its National Food Waste Recycling Service for businesses earlier in the year.

Ian Wakelin added: 'We all want to recycle more. It is initiatives like this which allow us to do so with minimum effect on our day-to-day lives at home or at work.'

Every year, the UK throws away around 15 million tonnes of food waste (source: Defra/WRAP) and it is thought that around half of this comes from businesses. Much of this food waste is currently sent to landfill sites where it breaks down into methane and carbon dioxide, both powerful greenhouse gases that contribute significantly to climate change.

A video clip of the new facility is available at:

www.biffa.co.uk/about-biffa/media-centre/videos.html.

Notes

- Biffa is a leading integrated waste management business providing collection, treatment, recycling and technology-driven energy generation services.
- Biffa promotes and delivers sustainable waste management solutions for commercial and municipal waste.

17 June 2011

- The above information is reprinted with kind permission from Biffa. Please visit http://www.biffa.co.uk/ for further information.

Cannock waste plant stench still causing misery

Residents who have suffered foul smells from a Cannock waste plant are not convinced the problem has gone away despite £800,000 of improvements.

Householders living in Newlands Park say they have not noticed any real difference since Biffa cleaned up its act – but believe the real test will come in the spring when temperatures rise. Nursery manager Karen Hayward, aged 49, who lives in Peregrine Way, half-a-mile away, said:

'The problem has not gone away, although it seems to have lessened.'

She added: 'I'm out at work most days so it's hard to tell for sure but when we do get it, it's very strong.'

The Environment Agency recorded around 20 reports of excessive odour seeping from the £24 million plant in November. Officials then gave Biffa four months to make amends and will review the situation in March.

'I'm dreading the summer,' said Amy Pagett, 26, who moved to the estate a few weeks ago.

The trainee hairdresser, of Sparrowhawk Way had been warned by neighbours about the stench from an anaerobic digester installed last year at the Poplars plant.

Michelle McPherson, a 38-year-old IT manager, said: 'I can't say we've noticed any difference lately.'

The mother-of-two added: 'The smell has been so bad that we were worried about the health risk to our children. The firm has assured us that there isn't any danger.' Not everyone is bothered by the odour, however. Construction manager David Simner, 33, of Pheasant Way, rarely notices it.

He said: 'It depends on which way the wind is blowing but we seem to escape the worst of it.' The facility processes up to 120,000 tonnes of food waste from homes and businesses every year.

Engineering and processing director Dr John Casey said the firm had implemented 'the best available techniques' at Cannock to keep odour leakage to a minimum.

21 December 2011

- The above information is reprinted with kind permission from *Express & Star*. Please visit www.expressandstar.com for further information.

Too Good To Waste!

Sustainable restaurants ask diners to take their left-overs home.

The Sustainable Restaurant Association (SRA) is encouraging restaurants to reduce the 600,000 tonnes of food we all waste every year, with Too Good to Waste doggy boxes, asking diners to take their left-overs home.

Too Good To Waste, the SRA's campaign to tackle food waste in restaurants, launched on 5 October, kicking off with a cook off in Clerkenwell.

The average restaurant produces nearly half a kilo of food waste per diner per meal and the industry as a whole throws out 600,000 tonnes of food waste every year – most of it filling up scarce landfill space. It's costing restaurants, diners and the environment a fortune! Food waste is a serious issue. For every meal eaten in a UK restaurant, nearly half a kilo of food is wasted – through preparation, spoilage and what's left behind on the plate.

As part of the solution, the SRA's campaign is asking restaurants and diners to 'think inside the box' by introducing them to the doggy box. The SRA wants to make it not just acceptable but positive for diners to ask to take food home, while at the same time raising consumer and industry awareness about the scale of restaurant food waste.

The campaign is backed by top chefs including:

⇨ Hugh Fearnley-Whittingstall

⇨ Thomasina Miers, Wahaca

⇨ Antony Worrall Thompson

⇨ Cyrus Todiwala, Café Spice Namasté

⇨ Bruno Loubet, Bistrot Bruno loubet

⇨ Silla Bjerrum, Feng Sushi

⇨ Henry Dimbleby, Leon

⇨ Sam Harrison, Sam's Brasserie

⇨ Anna Hansen, The Modern Pantry

⇨ Sriram Aylur, The Quilon

⇨ Lisa Drabble, co-owner Squid and Pear

So help the SRA squash the food waste scandal: show you're a lover not a leaver! Visit the Too Good to Waste website at: www.toogood-towaste.co.uk.

5 October 2011

⇨ The above information is reprinted with kind permission from Sustain. Please visit www.sustainweb.org for further information.

Gleaning Network UK

Where there's muck, there's brassica!

Gleaning Network UK is an exciting new initiative to save the thousands of tonnes of fresh fruit and vegetables that are wasted on UK farms every year.

Farmers across the country often have no choice but to leave tonnes of their crops unharvested, which gets ploughed back in the soil. These crops cannot reach the market either because they fail to meet the retail strict cosmetic standards or because of overproduction. At the same time, 5.8 million people suffer from deep poverty in the UK and cannot afford a decent diet, and this number is on the rise.

Gleaning Network UK coordinates teams of volunteers, local farmers and food redistribution charities in order to salvage this fresh, nutritious food and direct it to those that need it most.

Several tonnes of excellent British produce – enough for thousands of meals – have already been saved in Kent, Sussex, Lincolnshire, and new gleaning groups are being formed in Manchester and Bristol. Apples, kale, cauliflowers, red, white and savoy cabbages have been among the crops saved by the Gleaning Network so far.

The 'Feeding the 5000' campaign is appealing to farmers across the UK to join in and allow volunteer 'gleaners' to harvest their unwanted produce for charity.

What is gleaning?

Gleaning is an ancient practice, dating back to medieval times, sometimes called 'scrounging'. It involves collecting crops from farmers' fields after they have been harvested.

Get involved

Are you a farmer with surplus produce or a redistribution charity and want to become part of Gleaning Network UK? Are you an individual and want to become a modern-day gleaner?

⇨ The above information is reprinted with kind permission from Gleaning Network UK. Please visit www.feeding5k.org for further information.

Key facts

- The UK sends 44 million tonnes of waste to landfill every year. This generates methane emissions that is estimated to account for 3% of all UK emissions. (page 3)
- Every tonne of food and drink waste creates around four tonnes of CO_2. £12 billion of good food and drink are wasted in the UK every year. (page 3)
- English households recycle 40% of their waste, up from little more than 10% a decade ago. (page 3)
- Waste services cost the average household over £145 a year. (page 4)
- All existing landfill sites in East Anglia have a maximum of ten years of operation left. (page 5)
- Studies have shown that a ten-acre landfill site will leak between 0.2 and ten gallons of liquids a day through their plastic liner. (page 5)
- In 2011/12, 43% of waste collected by local authorities was sent for recycling, composting or re-use. 37% went to landfill. The highest rate achieved by a council was 69%. The lowest was 14%. (page 6)
- The European target for household recycling rates by 2020 is 50%. (page 6)
- At least £650 million worth of valuable materials are being thrown into landfill or burned in the UK each year. (page 7)
- Fly-tipping costs local authorities almost £74 million a year to clear up. (page 8)
- It currently costs between 60p and 80p to dispose of a tyre in the UK. To avoid paying this much, an individual dumped over 800,000 tyres at environmentally-sensitive locations in Essex, Norfolk, Yorkshire, Worcestershire and Lincolnshire. (page 12)
- Smokers' materials are the most frequently recorded litter type. Cigarette butts and other litter from smoking were found at 82% of sites surveyed in England. (page 13)
- 21% of bins surveyed were deemed below an acceptable standard of cleanliness. (page 13)
- Almost 50% of the total amount of food thrown away in the UK comes from our homes. We throw away 7.2 million tonnes of food and drink from our homes every year, which costs the average household £480 a year. (page 15)
- In the European Union alone, around 8.7 million tonnes of e-waste are thrown away each year, only 2.1 million tonnes, or 25%, is collected and treated. (page 21)
- The average UK family throws away six trees' worth of paper in their household bin each year. (page 22)
- Countries like Switzerland, Germany and The Netherlands recycle around 60% of their waste. (page 22)
- A survey of beaches in the UK found that there was an average density of 38 plastic bags for every km surveyed. (page 25)
- Approximate percentages of waste arising in the UK indicate that around 36% is from construction, 28% is from mining and quarrying and 24% is household or commercial waste. (page 32)
- Over 90% of the UK's energy supply is provided from fossil fuels. (page 33)
- Sweden recently ran out of rubbish and had to import 80,000 tonnes of refuse from elsewhere. (page 37)
- The average restaurant produces 600,000 tonnes of food waste every year. (page 39)
- 5.8 million people suffer from deep poverty in the UK and cannot afford a decent diet. (page 39)

Glossary

Agenda 21

An action plan from the United Nations which focuses on sustainable development. In terms of waste management, Agenda 21 emphasises the need to reduce the amount of waste we make and calls for more recycling and re-using of waste products.

Biodegradable waste

Materials that can be completely broken down naturally (e.g. by bacteria) in a reasonable amount of time. This includes organic materials such as food waste, paper waste and manure, which can be composted, as opposed to items such as plastic bottles that would take thousands of years to break down naturally.

Fly-tipping

Sometimes referred to as 'sneaky dumping' or 'dumping on the fly', fly-tipping is the illegal dumping of waste in inappropriate areas. People usually do this so they don't have to pay for bulky items to be collected and removed. Fly-tipping is unsightly and poses a threat to the environment and human health.

Incineration

A method of disposing of waste by burning it into ashes. Incineration reduces the amount of waste that is sent to a landfill and can even convert waste into energy. However, there are concerns about the environmental impact of incinerators (air pollution, toxic waste, etc.).

Landfill

A site where waste is buried in a hole in the ground. As the waste rots and breaks down, poisonous and contaminating by-products are produced, such as a polluting liquid known as leachate and landfill gases (mainly methane and carbon dioxide).

Litter

Rubbish that has been discarded and left lying around rather than disposed of properly. Littering is a crime and people can be fined on-the-spot up to £80.

Recycling

The process of turning waste into a new product. Recycling reduces the consumption of natural resources, saves energy and reduces the amount of waste sent to landfills.

The three Rs/Waste hierarchy

The three Rs of recycling are Reduce, Re-use and Recycle. This refers to reducing the amount of waste you make, re-using materials rather them throwing them away (for example, glass milk bottles delivered to your doorstep get used again) and recycling materials by breaking them down and remaking them into something else (plastic drink bottles could be melted down and be made into a plastic chair).

Throwaway/throw-away society

A society where rather than re-using or recycling something, people just throw it away. This is strongly influenced by consumerism, the increased consumption of goods.

Waste

Anything that is no longer of use and thrown away. Each year the UK generates approximately 290 million tonnes of waste, which has a damaging effect on the environment.

Waste Electrical and Electronic Equipment (WEEE) directive

This EU Directive aims to reduce the amount of waste from electrical and electronic equipment and increase its re-use, recovery and recycling. Electrical items cannot just be throw away, they have to be disposed of in a particular manner. This includes large and small household appliances (e.g. washing machine or kettles), lighting equipment, electric tools, toys, etc. Before the implementation of the WEEE directive, waste like this was simply thrown into the bin, but now producers and supplies of electrical goods need to offer their customers a way to dispose of their waste properly (known as 'takeback').

Zero waste

A plan to promote the idea of recycling and re-using materials rather than just disposing of them. The aim is to reduce the amount of waste sent to landfills.

Assignments

1. Do we have a moral duty to recycle our rubbish? Discuss this question with a partner and make notes. Feedback to the rest of the class.

2. Read *Government review of waste policy 2011* on page 3. Write a summary for your school paper. Try to make the article relevant to the pupils at your school.

3. Write a report entitled *Waste: past, present and future.* You should divide your report into three sections: section one should look at how waste was disposed of in the past, section two should explore what happens to our waste today and the final section should consider what might happen to our waste in the future. Your report should be no longer than four sides of A4.

4. Read *What is fly-tipping?* on pages 8 – 9. Choose one of the most commonly fly-tipped materials, listed on page 9, and research how they should be correctly disposed of. Then try to think of some innovative ways your chosen material could be recycled. Write a few short paragraphs explaining your findings and your ideas.

5. Go outside and conduct a survey of the various types of litter you see. Present your findings in a graph. Is there more of a certain type of litter? Why is this? How could you reduce the amount of litter in the area? Write some notes on your findings.

6. Read *How much food is wasted in total across the UK?* on page 14. For one week, keep a detailed diary of how much food you throw away. Write everything down and at the end of the week create a table to demonstrate how much of each type of food you have wasted. What could you have done to reduce this waste?

7. Research a charity or organisation that tackles the issue of food waste in the UK. Create a PowerPoint presentation, to be shown to your class, explaining what your chosen organisation does and encouraging people to get involved. For example, you could look at Food Cycle, The Gleaning Network or The Sustainable Restaurant Association.

8. What is e-waste? Design a leaflet explaining the concept of e-waste and highlighting the problems involved.

9. Read *20 facts about waste and recycling* on page 22. Select a couple different facts and design an illustrated poster about them.

10. Visit your local council's website and explore its policies on recycling. Create a factsheet that could be distributed to people in your local area, explaining the council's policy.

11. Do you think plastic carrier bags are a good idea or a bad idea? Explore both sides of the argument and present your findings.

12. You have just found out that there are plans to build a waste-to-energy plant/factory near your home town. After exploring the benefits and drawbacks of this development, decide whether you are pro- or con- and plan a campaign which supports your point of view. Your campaign could include newspaper articles, posters, signs, leaflets and even a website. Provide samples of at least three different campaign materials.

13. The Sustainable Restaurant Association is tackling food waste in restaurants by encouraging diners to take their left-overs home (see *Too Good to Waste!* on page 39). Imagine they have asked you to produce a short advert to help promote their cause. Be sure to include details of their 'doggy box scheme', as well as facts about food waste. You might even want to make use of the celebrities already involved in the project.

14. In Germany some vending machines act as 'reverse' bottle machines. People pay slightly more for their bottled drinks, but when their bottles are empty they can return them to the machine and get money back for recycling. Do you think this would encourage people to recycle more? Why or why not? What other ideas can you think of to encourage people to recycle everyday waste?

15. Write a list of items that you or your family have thrown away in the last couple of months. Try to think of some alternative uses for these things. For example, that old t-shirt you threw away could have been re-used and turned into a pillowcase.

16. List different types of waste management (e.g. incineration, landfills, recycling, zero waste, etc.) and create a list of pros and cons. Which method would you recommend?

Index

Acknowledgements

The publisher is grateful for permission to reproduce the following material.

While every care has been taken to trace and acknowledge copyright, the publisher tenders its apology for any accidental infringement or where copyright has proved untraceable. The publisher would be pleased to come to a suitable arrangement in any such case with the rightful owner.

Chapter 1: What is Waste?

The problem with waste © Waste Online 2013, *Government review of waste policy 2011* © Department for Environmental, Food & Rural Affairs 2013, *Agenda 21 and waste management* © Sustainable Environment 2013, *What happens to our waste?* © RecyclingExpert 2000 – 2013, *England recycles more than it landfills for first time* © Energy & Environmental Management 2013, *Call for urgent action over burning and throwing valuable materials in landfill* © 2013 Guardian News & Media Limited, *What is fly-tipping?* © Keep Britain Tidy 2013, *Cracking down on waste crime* © Environment Agency and database right, *How clean is England?* © Keep Britain Tidy 2013, *How much food is wasted in total across the UK?* © Love Food Hate Waste 2013, *UK families waste £270 a year on discarded food* © 2013 Guardian News & Media Limited, *Our biggest litter count yet!* © Keep Britain Tidy 2013, *EU exporting more waste, including hazardous waste* © European Environment Agency (EEA) 2012, *No more plastic waste to China?* © Recycling & Waste World 2013, *What is e-waste?* © E for Good 2013, *The Great Pacific Garbage Patch* © European Union, 1995 – 2012.

Chapter 2: Tackling waste

20 facts about waste and recycling © CB Environmental Ltd 2013, *Recycling etiquette* © Fubra Limited 2013, *Recycling myths* © Recycle for London 2013, *Break the Bag Habit!* © MCS 2013, *Facts about supermarket carrier bags* © Incpen 2013, *Plastic-eating fungi discovered* © 2013 First News, *Plastic bag ban triggers innovative asphalt* © Green Futures magazine, Forum for the Future, 2013, *Cleaning up India's waste: but what is the future for army of tip pickers?* © 2013 Guardian News and Media Limited, *Bin fines up to £1,000 continue* © Louise Gray/The Daily Telegraph, *Rewarding you for recycling* © The Royal Borough of Windsor and Maidenhead 2013, *Energy from waste* © Institution of Mechanical Engineers 2013, *Why oppose incineration?* © United Kingdom Without Incineration Network (UKWIN) 2013, *Sweden runs out of garbage, forced to import from Norway* © Mother Nature Network 2013, *UK's first 'super' food waste plant opens* © Biffa 2013, *Cannock waste plant stench still causing misery* © Express & Star 2013, *Too Good to Waste!* © Sustain 2013, *Gleaning Network UK* © Gleaning Network UK 2013.

Illustrations:

Pages 7, 27, Don Hatcher; page 37, Simon Kneebone; pages 12, 25, Angelo Madrid.

Images:

Cover and pages i & 1 © gorich, page 8 © Alan Stanton, pages 10 & 11 © Jackie Staines, page 15 © Ilya731, page 20 © John Evans, page 28 © Arial Da Silva Parreira, page 32 © Norbert Nagel.

Additional acknowledgements:

Editorial on behalf of Independence Educational Publishers by Cara Acred.

With thanks to the Independence team: Mary Chapman, Sandra Dennis, Christina Hughes, Jackie Staines and Jan Sunderland.

Cara Acred

Cambridge

May 2013